IF WAR COMES, HOW TO DEFEAT SADDAM HUSSEIN

by

Colonel Trevor N. Dupuy, USA, Ret.,
Curt Johnson, David L. Bongard, Arnold C. Dupuy

Editor: Paul F. Braim, PhD.

1991

HERO BOOKS

McLean, Virginia

Some other recent works by Trevor N. Dupuy

ENCYCLOPEDIA OF MILITARY HISTORY
UNDERSTANDING WAR: History and Theory of Combat
UNDERSTANDING DEFEAT
ELUSIVE VICTORY: The Arab Israeli Wars 1947 - 1974
FLAWED VICTORY: The Arab Israeli Conflict and the 1982 War in Lebanon
A GENIUS FOR WAR: The German Army and General Staff 1807 - 1945
EVOLUTION OF WEAPONS AND WARFARE
DICTIONARY OF MILITARY TERMS
NUMBERS, PREDICTIONS, AND WAR
ATTRITION: Forecasting Battle Casualties and Equipment Losses in Modern War

© 1991 by T. N. Dupuy

ISBN 0-915979-27-6

All rights reserved. No part of this book may be reproduced in any form or by any means, except for inclusion of brief quotations in a review, without permission in writing from the publisher.

Printed and bound in the USA

HERO BOOKS
1324 Kurtz Road
McLean, Virginia 22101
Phone   703-847-3917
Fax   703-827-2683

Table of Contents

List of Maps and Diagrams iv
Introduction v
Chapter 1:   Background of the Gulf Crisis 1
Chapter 2:   Peace or War? The Strategic Options 19
Chapter 3:   The Thorny Problem of Command 27
Chapter 4:   How to Fight: The Tactical/Operational Options 34
Chapter 5:   Option One: Operation "Colorado Springs" 53
Chapter 6:   Option Two: Operation "Bulldozer" 62
Chapter 7:   Option Three: Operation "Leavenworth" 70
Chapter 8:   Option Four: Operation "RazzleDazzle" 74
Chapter 9:   Jordanian Diversion 80
Chapter 10: The Sobering Effect of Logistics 85
Chapter 11: Option Five: Operation "Siege" 95
Chapter 12: The Decision 100
Chapter 13: After the War is Over 108
Appendix A: Chronology: Iraq and the Gulf Region, 1514-1991 116
Appendix B: The Desert Environment 123
Appendix C: Equipment and Weapons of the Forces 133
Appendix D: Organization of Opposing Forces 141
Appendix E: Relative Combat Effectiveness 153
Appendix F: Logistics in Historical Perspective 156
Bibliography 161
Index 163

List of Maps and Diagrams

Figure 1.	Eastern Ottoman Empire, 1914	3
Figure 2.	Middle East Region, 1990	7
Figure 3.	Iraq, Jordan, and Vicinity	11
Figure 4.	Kuwait	17
Figure 5.	UN Command Structure	30
Figure 6.	Estimate of the Situation	36
Figure 7.	Deployment Situation, Kuwait and Vicinity	38
Figure 8.	Numerical Force Comparison	40
Figure 9.	Technological Force Comparison	41
Figure 10.	Technological and Effectiveness Force Comparison	43
Figure 11.	Technological, Effectiveness, and Posture Force Comparison	44
Figure 12.	Basic Combat Maneuvers	48
Figure 13.	Operation "Bulldozer"	64
Figure 14.	Iraqi Triangular Defensive Position	66
Figure 15.	Operation "Leavenworth"	72
Figure 16.	Operation "RazzleDazzle"	76
Figure 17.	Supply Classes	87
Figure 18.	Supply Expenditure Per Day, in Tons	88
Figure 19.	United States Forces' Casualties from D-Day to D+40	104
Figure 20.	Iraqi Casualties	106
Figure 21.	UN Casualties' from D-Day to D+40	106
Figure 22.	Climate Time-Line (Kuwait and Southeastern Iraq)	131
Figure 23.	Unit Symbol Legend	140
Figure 24.	US Mechanized Division	140
Figure 25.	US Armored Division	142
Figure 26.	US Air Assault Division	142
Figure 27.	US Airborne Division	144
Figure 28.	US Marine Corps Division	144
Figure 29.	British 1st Armored Division	146
Figure 30.	French 6th Light Armored Division	146
Figure 31.	Egyptian Armored Division	148
Figure 32.	Egyptian Mechanized Infantry Division	148
Figure 33.	Syrian Armored Division	150
Figure 34.	Iraqi Armored Division	150
Figure 35.	Iraqi Mechanized Infantry Division	152
Figure 36.	Iraqi Infantry Division	152

INTRODUCTION

This book is an elaboration of testimony which I presented to the House Armed Services Committee on 13 December, 1990. I received considerable assistance from three of my colleagues—Curt Johnson, David L. Bongard, and Arnold C. Dupuy—in the preparation of that testimony, and even more in the formulation of this book, and so they are truly my co-authors, as shown on the title page. In fairness to them, however, I should say that they loyally adapted themselves to my concept and organization, and so should not be held responsible for shortcomings in the approach or the organization of the book.

The purpose of the book is not to forecast the outcome of the conflict which looms over the Persian Gulf region as this book is being written. Nor is it an effort to second-guess the commanders and staffs of our forces now deployed in that region as they plan for that conflict. Rather, it is to acquaint those who might be interested in the kinds of considerations with which those commanders and staffs must be concerned in the planning process. If war does come, it surely will not take the form of any of the various options, or of the scenarios related to those options, which are presented in this book. But, on the basis of our military knowledge and experience, as well as our extensive knowledge of relevant military history, the authors are confident that there will be considerable similarity between the actual events, should conflict occur, and the possibilities suggested in this book.

It is not our intent to make a case for why the United States should go to war. We devoutly hope that war will not be necessary. The service men and women of our armed forces are the most precious of our national resources, and their commitment in "harm's way" should only be made in pursuit of objectives vital to the interests of the United States.

War in the Gulf may not occur for the best of reasons: the obvious power of the forces committed in support of the UN Charter, and our deliberate and public preparation to employ those forces may cause Saddam Hussein, even at the final hour before hostilities, to abandon his ill-gotten gains and accept the dictates of the United Nations Resolutions. It will be clear from what we have written, however, that a partial withdrawal of Iraqi power from Kuwait should not be accepted as a solution to the confrontation. Sober judgment leads us to the conclusion that a war with the Iraqi dictator in the near future is better for the United States and the world than would be a compromise solution that rewards aggression. Such a "diplomatic compromise" solution will inevitably assure a later and more deadly war with a much more powerful Saddam Hussein. We believe that the current Middle East crisis is more closely parallel to the Czechoslovakia crisis of 1938 than is usually found in history, and we hope that the lesson of that crisis, which ended in the Munich Agreement, will be heeded.

The original title that I selected for this book was "How to Defeat Iraq." This did not seem fully consistent, however, with our hope that war might possibly be avoided. I could not help thinking, therefore, of a book that my father and George Fielding Eliot wrote in 1938, when war clouds hung heavy over Europe, and they wished to educate the American public about the military strengths, weaknesses, and problems of the potential belligerents. That book was entitled: *If War Comes*. Since the purpose of this book is much like that which my father and George Eliot wrote, about 53 years ago, I thought the same title might apply to it. Thus, I changed my prospective title to: "If War Comes, How to Defeat Iraq."

While the book was being written I had an opportunity to discuss it with a young lady who is intimately involved in the unfolding events in the Middle East. A graduate of the University of Amman, she is a Palestinian (whose home is in Jordan), the daughter of an old and respected friend. She is also the wife of an Iraqi, a member of the opposition to Saddam Hussein, who is currently underground somewhere in the Middle East. I told her that the tentative title of the book was "How We Will Defeat Iraq." She said to me, "You are not enemies of the Iraqi people; your enemy is Saddam Hussein." She is right, of course, and that is how I arrived at the final version of our title.

My colleagues and I are indebted to a number of people who have provided helpful comments and advice in the preparation of the book, and/or the congressional committee testimony that preceded it. In particular I want to thank Major General John E. Murray, USA, Retired; Major General Fred Haynes, USMC, Retired; Lt. Col. Peter J. Clark, USAF, Retired; Lt. Commander James T. Westwood, USN, Retired; Mr. Philip McDonnell; and Mr. Robert McQuie. I also received extremely valuable assistance from two relatively senior Pentagon staff officers (one Army and one Air Force), both of whom reviewed the entire text of my written testimony, and several chapters of this book, and provided helpful comments. I am sure they will appreciate my appreciation, but I suspect they would just as soon remain anonymous.

January 1991

T.N. Dupuy
McLean, Virginia

CHAPTER 1

Background of the Gulf Crisis

From time immemorial the land known today as Iraq has been the scene of conflict. Encompassing much of the ancient Fertile Crescent watered by the Tigris and Euphrates rivers, Iraq has been not only a strategic highway linking the Eastern Mediterranean lands with those of the Orient, but also the scene of frequent clashes between empires and great powers.

Iraq has seldom been master of its own destiny, and in the numerous conflicts that stud its history, it has more often than not been pawn or prize of other powers seeking regional hegemony. Until recent times most conflicts in the region were imperialistic in nature and involved Iraq because of its strategically important position. The discovery of vast oil deposits in the region beginning in 1907 added another element to the equation, and conflicts since have sprung from imperialistic/hegemonic motives as well as from a desire to protect or control sources of much of the world's most important strategic resource. It was thus that in 1990 a local quarrel involving people far away of whom we know nothing (to paraphrase Neville Chamberlain) became a critical threat to world peace, and—for only the second time in its history—the United Nations took decisive action to oppose by military force naked, unprovoked aggression.

The roots of the present Gulf crisis extend back centuries, drawing sustenance from several historical springs. Among these are: the dissolution of the Ottoman Turkish empire; Arab nationalism and the pan-Arab movement; the ancient antagonisms of the Shiite and Sunni Moslem sects and of Arabs and Persians (Iranians); the voluntary breakup of the British world empire, and the consequent political vacuum in many regions, including the

Persian Gulf; the Arab-Israeli conflict; the Palestinian question; and endemic great power rivalry in Middle East.

The Ottoman Legacy

From the mid-17th Century until the outbreak of World War I the Ottoman Turkish Empire was the preeminent power in Iraq and in the Middle East generally. The Porte's writ was law as far east as the Persian Gulf, where, in 1756, recent settlers of the northwestern coast, members of the formerly nomadic 'Anizah tribe, selected as their sheikh a member of the Sabah family. This was the beginning of the Sabah dynasty, rulers of the modern state of Kuwait. It was also the beginning of Kuwaiti autonomy, occasionally contested, but never seriously challenged, by the Turks, who had other, more important, fish to fry.

The Ottoman-Safavid Rivalry

Turkish hegemony in Iraq was established gradually in a series of wars with the Persian Safavid dynasty during 1514-1746, but was never seriously disputed after 1638. Shah Ismail I (1500-1524) founded the Safavid dynasty, which was established in Azerbaijan and western Persia, and battled the Ottomans in the west and the Uzbek Tartars in the northeast (Khorasan). Ismail and his successors led formidable, typically eastern armies composed principally of skilled horsemen armed with bow and lance. These men were no match for the relatively modern Ottoman armies, which had disciplined infantry armed with firearms (the famous *Janissaries*—the modern world's first regular infantry), cavalry not unlike that of the Persians, and artillery. As a result, in the 16th and 17th centuries, the Persians for the most part were beaten routinely.

Shah Abbas I ("the Great," 1587-1629) was able temporarily to reverse the trend. He reorganized and modernized the Persian army with the assistance of westerners, including the peripatetic English Sherley brothers, who left a remarkable account of their achievements. But Abbas—not himself a Safavid, but a Barlas Turk—was assassinated, and the Turks reasserted themselves in the anarchy that followed his death. The Ottoman-Persian frontier that was eventually agreed is largely that which at present separates Iraq and Iran.

In the Gulf both Persian and Ottoman influence was weak. The littoral sheikhdoms and city-states were devoted to trade, slaving, and piracy, and looked to the sea for their sustenance, not to the turbulent, decaying interior.

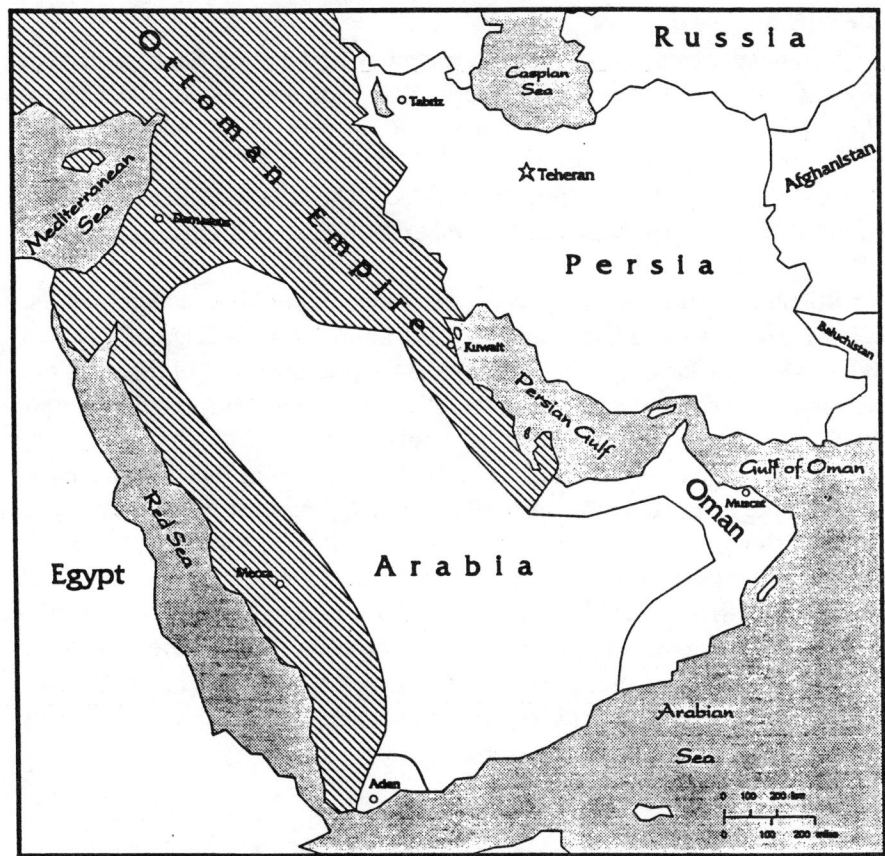

FIGURE 1. EASTERN OTTOMAN EMPIRE, 1914

The powers in the Gulf were Portugal, briefly, then England. By the mid-19th century the Gulf was virtually a British lake—albeit a backwater—and the British busied themselves by curbing piracy, improving navigation, and promoting trade between the Gulf and their imperial India. It was imperialism at its benign best and served further to distance the Gulf states from Baghdad and Constantinople.

When the Turks sought to advance their claim against Kuwait, as they did in 1898, the British provided protection to the sheikh and the Turks backed off. In 1899 Kuwait's status as a British protectorate was regularized, and in 1913, just prior to the outbreak of World War I, Turkey recognized the autonomous status of Kuwait, abandoning all pretense of sovereignty over the sheikhdom (and its Gulf neighbors Bahrain, Qatar, Muscat (Oman), and the Trucial Coast—now the United Arab Emirates).

World War I

Britain sent troops to occupy Basra and protect her oil fields in Persian Khuzistan after Turkey entered World War I on the side of the Central Powers in October 1914. After nearly four years of difficult and costly campaigning, and the loss of one small British army which surrendered at Kut (Al-Kūt) in April 1916, the British prevailed. They captured Baghdad on 11 March 1917 and took Mosul in November 1918, just days after an armistice officially ended hostilities.

British operations were severely constrained by terrain, weather, and the relatively narrow and difficult approaches to objectives in the Tigris-Euphrates valley. A major factor (for both sides) was the primitive state of the road/rail net and the limited capacity of alternative modes of transportation/supply (such as shallow draught steamers and rafts).

The British had an advantage in mobility conferred by their riverine navy, and had the additional advantage of a small air force that was unchallenged by Turkish air.

As part of the treaties that dismantled the old Ottoman Empire at the end of the war, Britain gained control of Iraq as a League of Nations mandate, endeavoring to govern it and prepare it to become a fully sovereign state. Iraq became an independent country in 1930, and was admitted to the League of Nations in 1932 with British sponsorship. The Gulf states were relatively unaffected by events of World War I.

World War II

The Anglo-Iraqi Treaty of Alliance and Mutual Support (1930) permitted the British to maintain military bases in independent Iraq and obligated the Iraqis to protect the bases from such dangers as attack by unruly tribesmen. The British also were permitted unimpeded transit for their troops between and among their various imperial holdings. The strategic importance of Iraq for the British at that time included: (a) defense of their sea line of communications Suez-India (the Persian Gulf flanks this major east-west line of communications); (b) defense of the Anglo-Persian oil fields; and (c) support of friendly Arabs. The outbreak of World War II in September 1939 and the generally pro-Axis sentiment of most Arabs posed a serious threat to British interests in the Middle East.

The Axis powers recognized the importance of Iraq in Middle East strategy and, as early as 1933, began to penetrate the Iraqi government and stir up anti-British sentiment in the region. One of their chief Quislings was Rashid Ali, a virulently anti-British nationalist who served as on-again, off-again prime minister during 1933-1941. His final opportunity came in early 1941 when British fortunes in the Middle East were at low ebb. Taking advantage of a weak regency, he and certain Iraqi officers conspired ("Golden Square conspiracy") to depose the pro-British monarch and drive the British from their bases in Iraq with the help of Axis troops.

The British, very weak in Iraq, were alive to the threat in early April but, hard-pressed on other fronts, were almost powerless to do anything about it. In May 1941 the Iraqis attacked the British bases at Basra and Habbaniya. The Indian Army was called on to reinforce Basra, and it appears that Basra was never seriously threatened.

The principal Iraqi effort was made against the British air base at Habbaniya, just 50 miles west of Baghdad, on the south bank of the Euphrates. Habbaniya was an RAF training base, home of No. 4 Flying Training School, 39 pilots, and a gaggle of obsolete aircraft, mostly biplanes. The Iraqis besieged Habbaniya during 2-6 May, utilizing a commanding position on a 200-foot high plateau just one mile from the British perimeter. The Iraqis had 5-9,000 men and 28 guns. The small, but relatively modern Iraqi air force (76 planes) was initially successful in the bombardment campaign that followed but, despite deficiencies in numbers and quality of aircraft, the British quickly gained the upper hand, destroying the Iraqi air force and dominating the air above Habbaniya.

The British conducted an active defense, making numerous air sorties. The Iraqi artillery was effectively suppressed by British air. Heavy bombers (Wellingtons) flying from Shaiba near Basra bombed the Iraqi lines. The Iraqis withdrew in a panic to Falluja on 6 May; on the same day an Iraqi reinforcing column came out from Baghdad. The two columns met on the road near Falluja, and the resulting traffic jam was pounded unmercifully by British air from Habbaniya. On that day alone the Iraqis had 1,000 casualties.

The promised Axis aid to Rashid materialized in the form of Italian and German military aircraft at Mosul; the RAF attacked the base and destroyed the Axis aircraft. From that point, the British were unchallenged in the air.

The Iraqis dug in at Falluja, 35 miles west of Baghdad, to defend the bridge over the Euphrates. The British at Habbaniya awaited reinforcements coming cross-desert from Palestine.

The first increment of the British reinforcement ("Habforce"—a motorized brigade group) arrived from Palestine on 18 May. On 19 May the British attacked at Falluja, attempting to seize the Euphrates' bridge intact. The Iraqi position was enveloped by means of a "flying bridge," and troops were air-landed in its rear on the Baghdad road. The British expected that the Iraqis would collapse, but this did not happen. The Iraqis finally succumbed to direct assault, and the British had to beat-off a counterattack from Baghdad three days later.

On 30 May the British arrived in the suburbs of Baghdad. On 31 May an armistice was negotiated and Rashid Ali fled to Persia.

In this one-month long crisis the Iraqis had the initiative briefly in the early stages, not because the British were surprised (they were not) but because the British were so weak in men and materiel that they were near powerless to prevent the Iraqis from acting. However, the resolute defense of Habbaniya took the wind out of the Iraqi's sails. The small RAF presence quickly gained air superiority, and was decisive. Air superiority permitted the British to apply overwhelming firepower at critical points and move and maneuver at will. Moreover, it destroyed the morale of the Iraqi forces, and was the decisive factor in the battles at Habbaniya and Falluja.

The Anglo-Iraqi conflict of 1941, though near-forgotten today, was one of World War II's decisive battles. Won in large part by a small band of RAF personnel against a determined enemy equipped with modern materiel, it thwarted Hitler's ambition to "wreck finally the English position between the Mediterranean and the Persian Gulf" (Führer Directive 30, Middle East, 23 May 1941).

Origins of the Modern Arab States Involved in the Gulf Crisis

All of the Arab states involved in the events of 1990-1991 attained their independence since the end of World War I, the more important of them during the interwar years. Since several had belonged to the Ottoman Turkish empire, they were League of Nations' mandates following the post-war treaties of Sèvres and Lausanne, which dismantled the empire.

Saudi Arabia

The birthplace of Islam and the site of its most sacred shrines, Saudi Arabia is an oil-rich, absolute monarchy that has been ruled by the royal House of Saud since its unification in 1932. Although strikingly conservative (some might say medieval) in its domestic and foreign policies, and a steadfast

FIGURE 2. MIDDLE EAST REGION, 1990

opponent of Israel, Saudi Arabia has been one of the United States' most loyal and dependable allies in the Middle East since the November 1943 Cairo Conference, when King Abd al-Aziz (the founder of Saudi Arabia, and known in the West as Ibn Saud) met with President Franklin D. Roosevelt.

The Iraqi invasion of Kuwait in August 1990 was so palpably a threat to Saudi Arabia that the United States was forced to move decisively to halt the aggression and attempt to restore the status quo ante.

Kuwait

Kuwait is relatively old as a geographical/political entity, but relatively young as a nation, having attained its independence from Great Britain on 19 June 1961. It is thus slightly older than most of the men making up the UN military force poised to restore its independence. However, it was an autonomous sheikhdom, under the present ruling family, as early as 1756.

In light of events in August 1990, it is worth noting that six days after Kuwaiti independence was proclaimed, Iraq claimed Kuwait and threatened invasion (25 June 1961). Kuwait appealed to Britain for aid, and Iraqi plans were foiled. When Kuwait was admitted to the Arab League on 20 July, Iraqi claims to the emirate were effectively quashed. This was the first—but not the last—Iraqi claim against Kuwaiti sovereignty.

It is undeniably true that, historically, Iraq and Kuwait (together with many other modern nations in the region) were often ruled by the same empire. However, Kuwait, although it has ethnic and cultural ties, as well as longstanding economic connections, to Iraq, is most closely linked to the maritime trading states of the Gulf (Oman, Bahrain, and the like). And, Iraq's claims on Kuwait have about as much legal-historical justification as would a United States claim to Canada, or an Australian claim to New Zealand, solely because in each case both countries were former parts of the British Empire.

Iraq

Modern Iraq includes ancient Mesopotamia (Greek for "the land between the [Tigris and Euphrates] rivers"), or Upper Iraq, north of Baghdad, and Babylonia, Lower Iraq, south of Baghdad. Sustained by the life-giving waters of the rivers, the region was "the cradle of civilization," familiar from history books and biblical associations. This region, together with the northern Nile Valley of Egypt, has been settled and civilized for at least 5,000 years. In this context, "civilized" implies no value judgments, but refers instead to the practice of intensive agriculture, supported by irrigation, and the presence of numerous urban centers.

Until the final collapse of the Ottoman Empire in 1918 Iraq was commonly called Mesopotamia and comprised four Turkish *vilayets*, or provinces, ruled by a governor-general.

Agriculture in the area certainly predates 3000 B.C., and goes back to at least to 7500 B.C. in the fringe of hill country just north of the Fertile Crescent proper. Irrigation, at least in the lower Tigris-Euphrates valley, is a more recent development, and appears to date to about 5000 B.C. The ancient cities of Sumer and Akkad (modern southeastern Iraq) date to the late fourth millennium B.C., about 5,000 years ago. This means that Iraq is one of the oldest continually settled areas on the face of the earth. Only the Nile valley is reliably considered to have been settled for so long. Iraq is also the site of the oldest evidence of written records, marked on clay tablets about 3100 B.C.

For most of the last 4,000 years, the area now covered by Iraq has been governed by a series of large empires. Some of these have been foreign, while others have been centered in Mesopotamia. These states included:
- Hammurabi's Babylonian empire (1720-1530 B.C.)
- Earlier (1350-1050 B.C.) and later (900-615 B.C.) Assyrian empires
- The New Babylonian Empire (625-539 B.C.)
- The Achaemenid Persian Empire (539-331 B.C.)
- Alexander the Great's Empire and its Seleucid successor-state (331-c. 150 B.C.)
- The Parthian Empire (c. 150 B.C.-227 A.D.)
- The Sassanid Persian Empire (227-640)
- The Ommayyad (661-750) and Abbassid (750-1258) Caliphates
- The Mongol Il-khan Empire (1258-1355)
- Tamerlane's Empire (1355-1405)
- The Safavid Persian Empire (c. 1500-1534)
- The Ottoman Turkish Empire (1534-1918)
- The British Empire (1918-1930)

The most important historic influences on Iraq (and the Middle East in general) have been Islam and colonialism/imperialism, together with their modern correlates: Arab nationalism and the pan-Arab movement.

The history of Iraq during World Wars I and II has been recounted above. The recent history of Iraq may be said to have begun with the overthrow of the British-established monarchy in a bloody revolution in 1958 (14 July). Subsequently, Iraq underwent numerous changes of government between 1958 and 1971. During that 13-year period, there were no fewer than seven coups (not all of them successful), and these were often bloody. The 1968 coup-cum-revolution brought the Ba'ath (Arab Renaissance) Party to power, and it has retained control of Iraqi politics since. Organized on the pattern of the Soviet Communist Party, the Iraqi Ba'ath Party is tightly centralized, closely controlled by its leaders, and strongly hierarchical.

During this same period, Iraq has been plagued by nearly continuous unrest and rebellion among its Kurdish population. Ethnically unrelated to the majority Arabs, the 3.25 million Kurds in Iraq (nearly one-fifth of the population) live in the highlands north and east of Mosul, an area which (unfortunately for prospects of Kurdish autonomy) contains much of Iraq's oil reserves.

Further complicating Iraqi political development has been that country's opposition to Israel. Iraq sent troops to support the Arab cause during the Israeli War for Independence (May 1948-January 1949), although these were withdrawn when the pro-Western government of Nuri es-Said took power in autumn 1948. Iraq was not involved in the 1956 war, but an Iraqi brigade sent to assist Jordan arrived too late to see action during the Six-Day War in June 1967. Iraqi military participation was limited to aerial combat which cost the Iraqi Air Force 12 planes. Iraq sent two divisions to support Syria during the 1973 October War with Israel; the performance of Iraqi troops in that war was considered poor by both Arab allies and Israeli enemies.

Saddam Hussein, the chief protagonist in the present conflict, assumed the reins of government in Iraq on 16 July 1979. Within a year, he plunged his country into war with Iran—a war that was to last eight years and cost hundreds of thousands of lives.

Syria

Named after the ancient Roman province, Syria was part of the Ottoman Empire from 1516 until late 1918, when it was entered by Allied troops at the end of the Palestine Campaign. A French mandate under the League of Nations (1920-1941), Syria was taken from Vichy French troops by British and Free French forces in July 1941. Nominally independent from 1943, in March 1945 Syria was invited to participate in the United Nations conference at San Francisco. A short time later Allied troops evacuated the country; on 17 April Syria declared its independence.

The first quarter-century of Syrian independence (1945-1970) was turbulent and marked by many coups and government changes. The Arab Socialist Resurrection (Ba'ath) Party (in various manifestations) has ruled since 1963, and former air force general, President Hafez al-Assad, one of the Middle East's wiliest politicians, has held unchallenged power since November 1970.

An implacable foe of its neighbor Israel, a long-time Soviet client state, and a sanctuary for vicious anti-Western terrorists, Syria is the "strange bedfellow" of the US-dominated anti-Iraqi coalition. However, Syria has also been an intractable enemy of the rival Iraqi Baathist regime and steadfastly supported Iran during the Gulf War. Thus, Syrian participation in the alliance is not inconsistent with either Syrian or US policy.

Jordan

The one Arab state apparently sympathetic to Iraq is Jordan (the Hashemite Kingdom of Jordan), ruled by King Hussein since 1953. A constitutional monarchy with a large Palestinian population and frequently hostile to Syria, Jordan provided much aid to Saddam's regime during the Iran-Iraq War and has remained precariously aligned with Saddam since. Most of the Palestinians in Jordan are supporters of Saddam Hussein because they see him as a

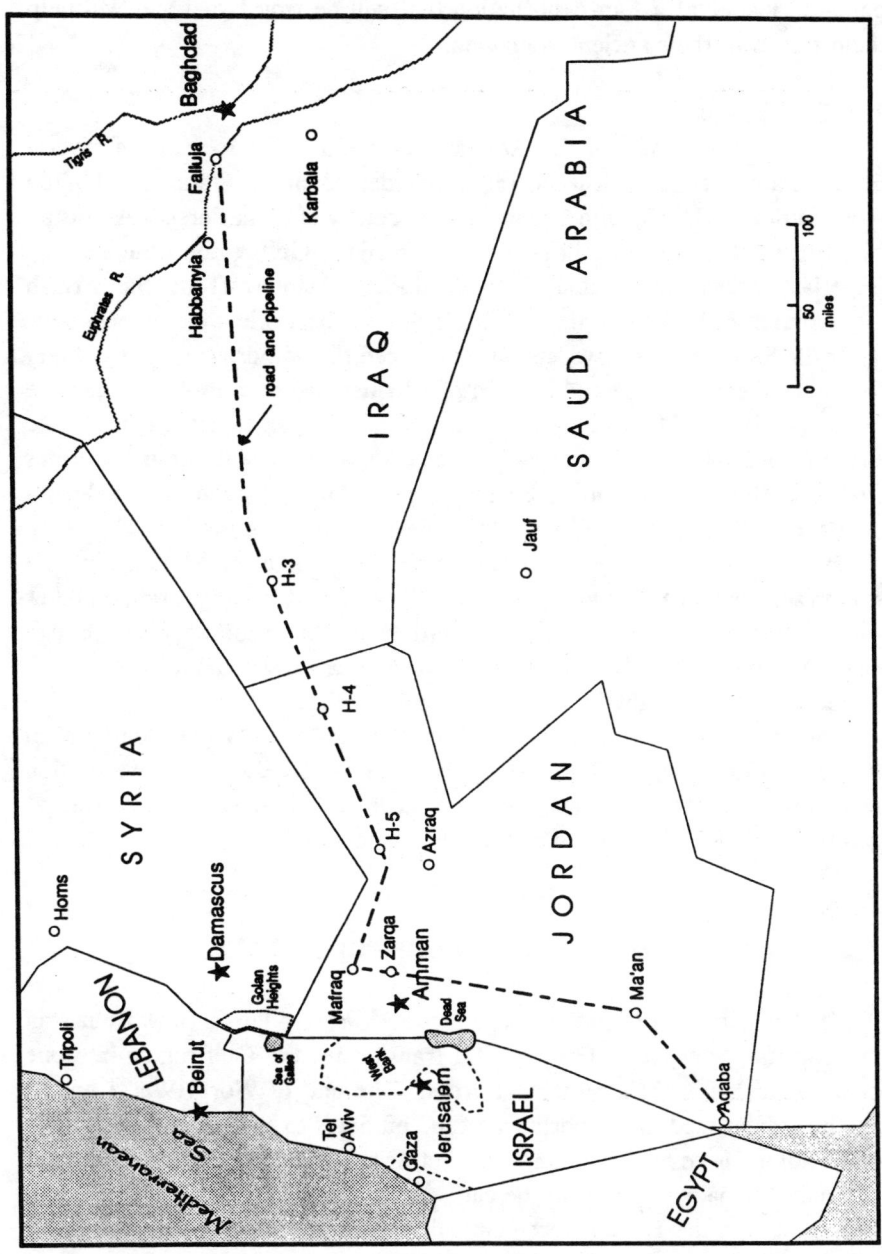

FIGURE 3. IRAQ, JORDAN, AND VICINITY

potential leader of a Pan-Arab nation that will destroy Israel and will help them return to their ancient homeland.

The Gulf States

The northeast coast of the Arabian Peninsula is the location of several littoral states. Besides Kuwait, these include: Bahrain, Qatar, the United Arab Emirates (UAE), and Oman. The discovery of oil in Persia's Arabistan by a British concessionaire in 1901 transformed the Gulf region from a sleepy imperial backwater, dominated (with studied disinterest) by the British Government of India, into the bustling center of the world-wide oil economy.

In 1968, the British government announced the withdrawal of its military forces from east of Suez and the end of a British presence in Arabia and the Persian Gulf by 1971. As a result of this decision, in part taken to concentrate Britain's military capabilities within NATO, the United Arab Emirates (formerly the Trucial States), Bahrain, and Qatar all gained their independence in 1971, and the Gulf states were left to their own devices.

Saudi Arabia, the largest state on the Arabian peninsula, soon acquired a role as spokesman and leader for the smaller countries nearby. Iraq, together with the other Arab nations, felt no particular debt to the Saudis, although guardianship of the Moslem holy city of Mecca provided the Saudis with important moral weight for their views and policies.

The Gulf states plus Saudi Arabia formed the Gulf Cooperation Council (GCC) in 1981, early in the Iran-Iraq War, to promote economic cooperation and provide a framework for collective security. The Iraqi rape of Kuwait galvanized the GCC states to oppose the aggression.

The Growing Importance of Oil

The exploitation of oil which began in southwestern Persia (now Khuzistan or Arabistan province of Iran) in 1901 transformed the Gulf region into one of the British Empire's most vital areas. The end of World War I had, if anything, increased the importance of the oil fields to Britian, although none of the adjoining countries became outright British colonies. The establishment of Saudi Arabia, completed in the early 1930s, changed the political equation in Arabia, and the British were compelled to come to terms with Ibn Saud. Oil prospecting during the 1920s and 1930s had little immediate effect, although oil production began in Iraq (reaching significant levels by the mid-1930s) and expanded in Iran.

Ibn Saud granted the first oil concessions in Saudi Arabia in 1933, and production began in 1937. This activity was highly successful and profitable, and led to the creation of the Arab-American Oil Company (ARAMCO) in 1943. By 1945, total production in the Gulf area amounted to 27.5 million metric tons annually, 17 million from Iran, 3 million from Saudi Arabia, and the rest from Iraq. Ten years later total production had risen to over 157 million tons; Iran's production was virtually unchanged, while Saudi production had risen to 47.5 million tons and Iraqi production to 33.5 million. Kuwait, which produced no significant oil in 1945, produced 54.5 million metric tons in 1955, and Qatar produced 5.5 million.

Oil production continued to expand during the 1960s and 1970s, with the Trucial States (now the United Arab Emirates) beginning production in the late 1950s, and Oman starting in the early 1970s. By the late 1980s, the states surrounding the Persian Gulf controlled about 51 percent of the world's oil reserves, and production from the Gulf states totalled nearly 778 million metric tons in 1988. The dependence of industrialized nations on oil has made the Gulf an area of unsurpassed global economic and political importance.

The importance of the Gulf for the world economy was marked by two major developments. The first was the Arab oil embargo imposed on the US and Western Europe during the latter stages of the October 1973 Arab-Israeli War, which definitively demonstrated the economic power of the Gulf states. The second development, which occurred more gradually, involved the creation of major financial and trading networks in the Gulf, and the transformation of traditional Bedouin sheikhdoms into small states with all the physical trappings of the modern world. The Gulf states employed their oil riches to create well-equipped (if generally small) armed forces, and to provide their citizens with an impressive array of services and facilities, including schools, clinics and hospitals, roads, and other manifestations of Western-type affluence.

The Emergence of Israel and the Arab-Israeli Wars

The startling development of the oil-rich Arab states of the Gulf has been paralleled by the equally impressive emergence of the Jewish state of Israel, which was formed out of the British League of Nations mandate of Palestine and attained independence in 1948. Literally "born in battle," Israel has fought four bitter wars for survival against coalitions of her Arab neighbors (see Appendix A). Israel, however, has achieved *rapprochement* with Egypt

(Camp David Accords, 1979), and despite the troublesome internal security problem posed by the continuing Palestinian *intifada*, is militarily strong enough impose her will on any and (probably) all of her immediate neighbors.

In the present crisis Saddam Hussein has threatened Israel and sought to link the Gulf question with Israeli occupation of territories conquered from Jordan and Syria in the 1967 Six-Day War. Although few find the linkage credible (and most understand Saddam's motivation), the possibility of an Iraqi attack on Israel, probably with long-range SCUD missiles, has created a threat to the cohesiveness of the UN coalition. It has also given him a base of political popularity in many Arab countries.

The Iran-Iraq (Gulf) War

The Islamic revolution in Iran (February 1979) brought a new factor into play in the political structure of the Persian Gulf states. The revolutionary regime which took power in Teheran was theocratic, militant and aggressive. Iraq, whose controlling Ba'ath Party was led by President Saddam Hussein, saw an opportunity. Iraq could, by striking while Iran was weak and sunk in political chaos, wrest concessions on the Shatt al-Arab, thereby not only improving Iraq's strategic position but increasing the stature of Saddam Hussein and his country in the Arab world. Hussein was also concerned about the prospect of the Iranian Shiite revolution spreading to Iraqi Shiites, who made up just over half of the Iraqi population.

Although Iraq was undeniably the aggressor and so bears immediate responsibility for the start of the war in September 1980, the fear which the Iranian revolutionary regime created among neighboring Arab states did much to make the Iraqi action more palatable to its neighbors. The creation of the Gulf Cooperation Council in 1981 was one indication of Arab mistrust and suspicion of the Iranian Khomeini regime. Another indication was provided by the extensive financial support given by several other Arab states to Saddam's regime in Iraq to enable it to keep fighting. Ironically, Kuwait was one of the major participants in this effort, no doubt because it lay closest to the battle lines and so had the most to lose if things went badly for Iraq.

Despite previous bad relations, both Egypt and Jordan came to the aid of Iraq. Jordan provided critical outlets for Iraqi oil (piped across the desert to the safe port of Aqaba on the Red Sea), and provided a brigade of several thousand volunteer soldiers. Egypt, despite Iraq's moves to bar it from the Arab League in the aftermath of the Camp David treaty, also lent Saddam

Hussein considerable support and assistance, including technicians and some materiel.

Saddam Hussein, who had pursued a cautious and essentially defensive strategy from 1981 to 1986, was compelled by the loss of Faw and the threat to Basra to consider something new after autumn 1986. The tremendous expansion of the Iraqi Army (infantry divisions increased from 10 to 30 during 1986-1987) is indicative of the effort put into the war effort, and a string of impressive Iraqi victories in summer 1988, achieved largely through superior numbers and firepower, compelled Iran to accept a cease-fire.

The Post-War Situation and Genesis of the Gulf Crisis

The relatively successful conclusion to the war from the standpoint of Iraq, together with the means by which it was achieved (an army totalling nearly one million men, with 5,000 tanks, 7,000 other armored combat vehicles, and some 3,500 artillery pieces) created the basis for a fundamental realignment of power in the Gulf region. With the military machine it had assembled, and with Iran too battered, bankrupt, and war-weary to cause much trouble for several years at least, Saddam Hussein was in a commanding position.

Limiting the rosy picture for Iraq was the issue of what the war had cost. Iraq's foreign debt, owed both to the Gulf states and to an assortment of European nations including France, Italy, and West Germany, was immense. The oil industry had been badly damaged by the fighting around Faw and Basra, and the economy had been badly hurt. Because of these considerations, most Western observers expected Saddam Hussein to spend several years attending to Iraq's economic recovery and the rebuilding of her war-damaged infrastructure.

Past Iraqi behavior, especially when it has perceived that its rivals are weak or divided, might have warned against such an evaluation. With hindsight, Iraq's invasion and occupation of Kuwait, following on several weeks of half-hearted negotiations and political maneuvering, was clearly a move consistent with Iraqi policy, even if it served to alienate countries which had recently been allies. Complicating and amplifying these causes of Iraqi action was Saddam Hussein's evident desire to secure concrete gains from Iraq's military machine, and his desire to play the role of "Pan-Arab Leader," last performed by Gamal Abdel Nasser of Egypt in the 1950s and 1960s. Having disposed of traditional rival Iran, at least for the time being, the action against Kuwait has secured Iraq's immediate—and long-desired—territorial goals in the region.

Iraq's 2 August 1990 invasion and annexation of Kuwait was condemned that same day by 14 of the UN Security Council's 15 members (Yemen did not participate) in Resolution 660. Almost immediately, as well, US forces were set in motion for the Persian Gulf and Saudi Arabia's northeast desert, interposing themselves between Saddam's forces in Kuwait and likely objectives of any possible Iraqi follow-on operation against Saudi Arabia (9 August). This was the genesis of Operation "Desert Shield." On 11 August Egyptian and Moroccan troops were deployed to Saudi Arabia, the first Arab elements of the multinational force opposed to Iraq to arrive in the theater.

Security Council Resolution 660 was followed in rapid succession by other resolutions which, among other things, imposed sanctions on Iraq (Resolution 661, 6 August), condemned Iraq's treatment of foreign nationals and diplomats, and demanded that Iraq cease its mistreatment of Kuwaiti and other nationals trapped in Kuwait. On 29 October, UN Resolution 674 warned Iraq that it was liable for damage done during the occupation, thus establishing a legal basis for reparations. On 29 November, in Resolution 678, the UN authorized the use of "all possible means" (i.e., military force) to implement its resolutions, setting 15 January 1991 as a deadline for Iraqi withdrawal from Kuwait and compliance with other resolutions.

The UN resolutions echoed the position of the US government, made plain on many occasions by the President and Administration spokesmen. Further, the President has repeatedly warned that (in an allusion to the Vietnam War) he would "never, ever agree to a halfway effort."

The historical record of the Gulf States, and especially that of Iraq and Kuwait, cannot provide a complete picture of the motivations and intentions of the major players in the current crisis. However, it is frankly impossible to begin to comprehend the current situation there, and in the Arab world at large, without some idea of what has passed before. Saddam Hussein and Iraq are not motivated solely by opportunistic greed, although that has certainly played a role in their evaluations, but by consistent long-term policy goals.

The political mandate of the Ba'ath Party and the residue of pan-Arab idealism cannot be ignored either; these have undoubtedly figured in the play made by Saddam Hussein for the hearts and minds of the Arab masses abroad.

To place in perspective the Robin Hood-like appeal of poor Iraq against rich and decadently westernized Kuwait, it is worthwhile to consider the situation at the end of World War I. In those days Kuwait was impoverished, its inhabitants making a marginal living from pearl-fishing and the coastal trade.

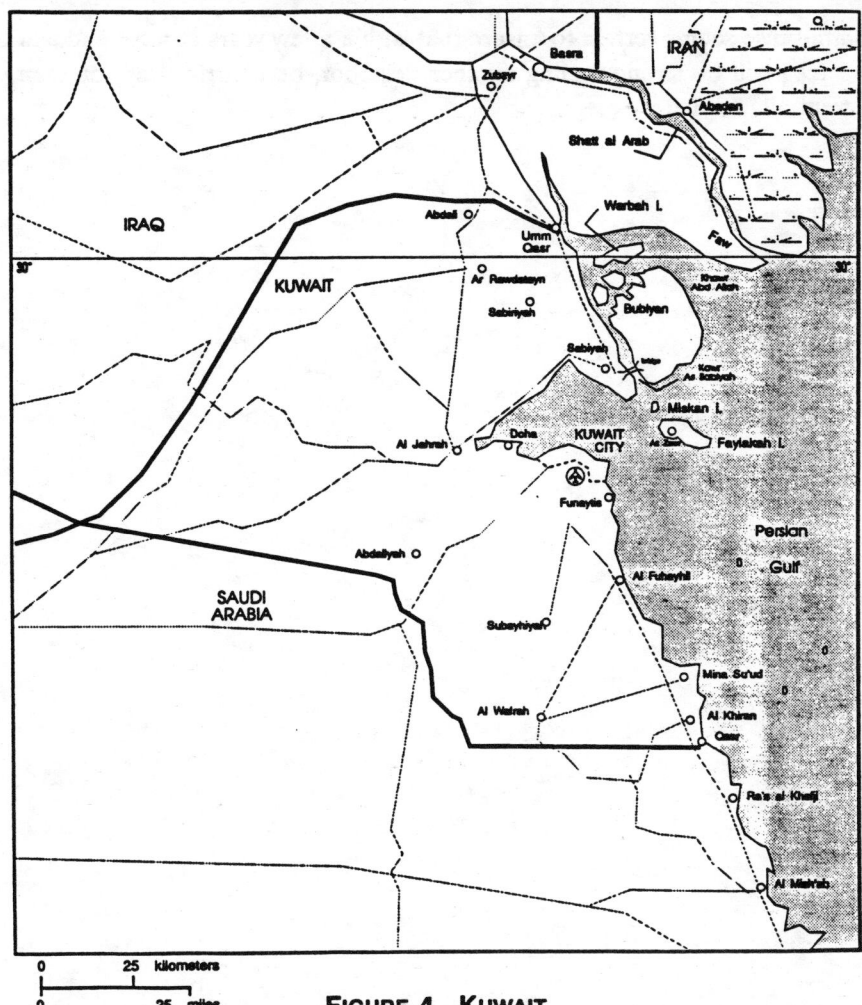

FIGURE 4. KUWAIT

Iraq was then a wealthy country, blessed with ample fertile land and a large population. Indeed, this pattern had prevailed throughout all recorded history, up to the opening of Kuwait's first oil well in 1946.

As a note of warning, any settlement of the current crisis negotiated with Saddam Hussein which leaves the Iraqi military machine intact will almost certainly ensure the return of war to the Gulf sometime in the next few years. Saddam's ambition for himself and his nation will not let him rest content with what he has; he desires a major leadership position in the Arab world, and a role as a major regional power for Iraq. While it is possible for the US and its UN compatriots to secure a peaceful settlement to this crisis, doing so by compromise seems certain to ensure that within a few years at most Saddam and Iraq will be strong-arming another neighbor, be it Syria, Iran, or even current ally Jordan.

CHAPTER 2

Peace or War? The Strategic Options

The commitment of large American military forces in the Persian Gulf region requires the United States to consider two very different kinds of options. First is the choice between the strategic options of peace or war. Second, should it come to war, is the choice between a veritable multitude of military operational or tactical options in accomplishing our objectives. In both cases, interestingly, the choice is virtually in the hands of the United States, and cannot be greatly influenced either by our enemies or our allies.

Let's look first at the strategic option of peace, as opposed to the option of war.

The Munich Parallel

"How horrible, fantastic, incredible, it is that we should be digging trenches and trying on gas-masks here because of a quarrel in a faraway country between people of whom we know nothing." Thus spoke Prime Minister Neville Chamberlain, in a radio address to the British people, 27 September, 1938 as Europe teetered on the brink of war because of the demands of German dictator Adolf Hitler for cession of Czechoslovak territory to Germany. Two days later Chamberlain and Prime Minister Edouard Daladier of France met in Munich with Hitler and his Fascist ally, Benito Mussolini of Italy. Chamberlain and Daladier avoided war by giving in to Hitler's demands. The Munich Agreement, awarding part of Czechoslovakia to Germany, was signed early on 30 September. That evening, after returning to London, Chamberlain announced to his cheering people that the Agreement was "peace with honor . . . peace in our time." Eleven months and three days later Great Britain was engaged in a desperate war for survival against Hitler's Germany.

In fact, the signature of the Munich Agreement virtually assured the outbreak of World War II. It is now known that if the Anglo-French allies had not capitulated at Munich, the Chief of the German General Staff, General Ludwig Beck, was prepared to overthrow Hitler. Thus, by pusillanimity, and the rewarding of aggression, Chamberlain and Daladier avoided a war in September 1938, but triggered the outbreak of a far deadlier war in September 1939, with a weakened coalition, against a much stronger enemy; a war in which France was destroyed, and Britain was driven to the brink of destruction.

Can Saddam Hussein, who has been called a "two-bit dictator," of a country with less than 20 million population, be compared either in power or malevolence with the Fuhrer of Germany in 1938?

Indeed, he can! If Saddam faces down the United States and the United Nations, even if he doesn't retain one square inch of Kuwait, unpunished for the brutal pillage and destruction of a sovereign member of the UN (and of the Arab League), retaining the most powerful military establishment in the Middle East, amply equipped with chemical and biological weapons, on the verge (in a few months or a few years) of achieving a nuclear weapons, capability, he will be the leader of a united (at least temporarily) Arab nation of more than 80 million people, and controlling more than half of the world's most important energy resource: oil. Think of how we would then have to fight him, as we try to rescue survivors of a new Holocaust!

There are powerful and persistent voices being raised to halt our slow and deliberate preparations for war, clamoring either to give the sanctions more time to work, or to reach a negotiated compromise solution with Saddam Hussein. These are voices of patriotic people of good will—as were those who cheered Chamberlain and Daladier in London and Paris the evening of 30 September, 1938. And if their shouts against war now prevail, they will be pushing us just as inevitably toward a far deadlier war, as did their kindred spirits in Britain and France half a century ago.

Other Parallels

For those who find the leap between Munich and Kuwait—or between Hitler and Saddam—too great for relevance, they might see the parallel in two other pre-World War II historical incidents.

On 5 December, 1934 an insignificant armed clash took place between Italian troops and an Ethiopian military contingent in the Horn of Africa at a place called Ualual. This was in an area claimed by Ethiopia as its sovereign territory, and claimed by Italy as a part of the colony of Italian Somaliland. Dictator Benito Mussolini threatened war unless Ethiopia apologized and paid reparations. The League of Nations attempted arbitration, while Great Britain and France endeavored to seek a diplomatic solution with Italy. Both efforts failed, because Mussolini saw an opportunity to create an Italian colonial empire in East Africa by the conquest of Ethiopia, and to avenge the disastrous defeat which Ethiopians had imposed upon an Italian army at Adua, 1 March, 1896.

On 3 October, 1935, Italian troops invaded Ethiopia from Eritrea in the northwest and Italian Somaliland to the east. After a surprisingly slow advance, the Italian Army—which did not greatly distinguish itself, but had the benefit of unopposed air support and the use of poison gas—occupied Addis Ababa, the Ethiopian capital, on 5 May, 1936. Emperor Haile Selassie fled to Europe, where he vainly pleaded for support and assistance from friendly governments and from the League of Nations. His impassioned and memorable speech to the League on 30 June, 1936, received polite applause, but accomplished little else. The League did impose sanctions on Italy, and considerable tension arose between Great Britain and Italy. But the League members could not agree upon including fuel oil in the sanctions, and Britain decided that Ethiopia was not important enough to risk war. The sanctions had little effect, and were soon ignored.

Had Britain—which had the support of France, Yugoslavia, Greece, and Turkey—been willing to risk war by supporting full sanctions against Italy, we do not know if Mussolini would have gone to war, or not. Had he done so, there can be no doubt as to the eventual outcome of such a war, and Munich probably would never have occurred. But had other events gone similarly, the elimination of the Italian end of the Fascist Axis would probably have assured Britain and France of an early victory, instead of disastrous defeat, in a war with Hitler's Germany.

Then there is the case of the Japanese annexation of Manchuria from China in 1932. As a legacy of past turmoil in China, Japan had the right to station troops in southern Manchuria, near the Japanese colonies of Port Arthur (Kwantung Peninsula) and Korea. (A similar right was exercised by the United States, which had a small infantry regiment at Tientsin.) During maneuvers of Japanese troops in southern Manchuria on 18 September 1931, a small explosion on a railroad (later proven to have been detonated by the Japanese themselves) gave a pretext for Japanese occupation of the vast region of Manchuria. They then created the puppet empire of Manchukuo, which was declared independent of China. (Viewers of the recent motion picture, "The Last Emperor," have at least been reminded of these events.)

This naked Japanese aggression was condemned by the League of Nations in general, and by the United States in particular. China declared a boycott against Japan, which was so effective that Japan occupied the Chinese metropolis of Shanghai and, after a long and bloody battle, forced China to end the boycott.

The rape of Manchuria is perhaps not so clear a parallel with the rape of Kuwait as were the subsequent rapes of Ethiopia and Czechoslovakia. The only nations in the world which might—collectively—have been in a position to force Japan to back down were the United States and Great Britain. But the isolationist sentiment in this nation would not have supported any real punitive action, and the lack of willpower in Great Britain at that time in history was soon demonstrated in the Ethiopian crisis. In retrospect, however, it is easy to see that the Japanese seizure of Manchuria set in motion the events that led ineluctably to World War II in the Pacific. And Japan's defiance of the League of Nations was a useful example to Mussolini with respect to Ethiopia, and Hitler with respect to Czechoslovakia. Japan could have been—and probably should have been—stopped. The cost could have been substantial, but certainly far less than that of World War II in the Pacific and Asia.

Let the Sanctions Work

The US policy which led the UN Security Council to present the 15 January ultimatum to Iraq has been strongly attacked in this country and abroad, but apparently by a relatively small segment of public opinion. Why, these people ask, shouldn't we try to make the sanctions work, so as to force Saddam to give in, rather than threatening war before we know whether the sanctions will or will not eventually be effective? And why, when he gives in, shouldn't we negotiate a reasonable, compromise peace with him? In fact, many of these people believe, we don't even need to wait for the sanctions to work; we can defuze the crisis, and get our young men and women back home from the Gulf, if we negotiate now. Those who offer any of these arguments would be indignant if they were accused of rewarding aggression. To them - this is of no significance, in comparison with the urgent matter of saving the world from war. It is peace at any price.

The only possible way that sanctions can work—other than to lead to negotiations which, *ipso facto*, rewards aggression—is to act as though we are convinced that sanctions will not work. If we are perceived to be clearly and unwaveringly determined to use force to achieve the aims of the United Nations resolutions, Saddam Hussein may recognize that, unless he surrenders, he will lose a war in which he will be destroyed, because his disillusioned people will kill him if he survives the conflict. He just might accept terms that will prevent him from repeating aggression, for the simple reason that while there is life there is hope. But he is not getting a message that tells him that he has the choice between peaceful defeat or destruction in war. He is getting the message of the people who are appealing for delays so that sanctions will work.

Which means that the sanctions will never work. Saddam's regime can continue as it is, without war, for at least a year, and maybe longer. The pressures to destroy the current UN coalition (including internal dissension in the United States) will lead to the collapse of the coalition in considerably less than a year. Thus, if he believes we will not go to war, Saddam Hussein can (and will) continue his bluff past 15 January. If we do not then call his bluff, he will win. He will keep Kuwait, he will retain his current overwhelming (in comparison to his neighbors) military strength and—after the embargo breaks down, as in inevitably will—he can continue his nuclear weapons, development program.

During that year—or during the portion of a year that the UN coalition holds together—there will be some suffering in Iraq. But it will be suffering in the civilian population, not the armed forces. Saddam has demonstrated that he will exploit with propaganda any instances of apparent, or real, suffering among his people. Pictures of starving Iraqi children on American television screens will contribute to the breakdown of American will to carry out the UN objectives. Thus, the sanctions will prove to be self-defeating.

But, one might ask, isn't the embargo preventing the Iraqi military machine from getting critical spare parts that it needs? Won't many of Saddam's fearsome weapons become useless if we wait for the sanctions to work? Yes, they will. But what is the significance of that if we are determined not to fight, and so he doesn't have to use the weapons?

There are, of course, two pragmatic arguments against the United States going to war in the current Gulf Crisis.

Should We Fight for Cheap Oil and Undemocratic Regimes?

One of these arguments is usually posed as a question: Why should American lives be sacrificed for cheap oil? The similar, but different, argument is also based upon a question: Why should American lives be sacrificed either to reinstate or to protect undemocratic regimes?

Both of these questions are loaded, and both are specious.

The significance of oil and oil economics is undeniable in the current crisis. But it is not a question of price. The issue is the threat that is posed to the world if more than half of the world's proven reserves of its greatest energy source were to be in the hands of a demonstrably brutal and ruthless dictator.

As to the internal political structures of the Kingdom of Saudi Arabia and the Emirate of Kuwait, these are totally irrelevant to the presence of US forces in the Gulf region. Two of our allies in World War II were the ruthless Communist dictatorship of the Soviet Union and the oligarchic semi-dictatorial government of the Republic of China. These countries were fighting on our side because they had been the victims of aggression by at least equally-ruthless dictatorial governments, which were also our enemies.

The United States is dedicated to law and order, both at home and in the world. A crime is just as heinous if committed against someone we do not like, or who goes to a different church, as it is when committed against a respected neighbor or beloved family member. What Saddam Hussein did to the legitimate government of Kuwait, and to its people, is at least as unforgivable as the crime that Hitler did to democratic Czechoslovakia and its people 52 years ago. It is as reprehensible as the crimes that the Japanese military oligarchy did against the non-democratic government of China, and that Mussolini did to the non-democratic government of Ethiopia. It broke international law, and it violated a solemn treaty to which Iraq was a party. It *must* not go unpunished. The crimes in 1931, 1935, and 1938 were initially condoned by weak-willed democracies. Let is not be said that the 1990 crime has been condoned by a weak-willed United States.

Obviously we cannot be a world policeman, risking precious American lives to combat agression wherever and whenever it occurs. But we should have no hesitation in asserting our righteousness in opposing aggression where it directly affects our national interests, as is the case with the Iraqi invasion of Kuwait.

Destroying Iraq Will Create a Destabilizing Vacuum

There is one final argument that can be used to support action—or inaction—in breaking the power of Saddam Hussein's militaristic regime. This is that if we punish Iraq too severely, a dangerous vacuum will be left in the Middle East, which will be filled by some equally unpleasant, opportunistic nation such as Assad's Syria or the unsavory theocracy of Iran and its ayatollahs. This argument is offered seriously, but it is hard to take it seriously. Equally unlikely is the possibility of an "unholy alliance" between Iran and Syria, once their common enemy, Saddam Hussein, has been eliminated from power. It is difficult to see how any outcome of the crisis, even if it comes to a war which totally destroys the military strength of Iraq, could be more unbalanced, or more unstable than the situation was after Iraq emerged semi-victorious from the Iran-Iraq War. Neither Iran nor Syria could be more dangerous to peace and stability than has been the regime of Saddam Hussein in Iraq, which gobbled up Kuwait, and was almost certainly ready to gobble up Saudi Arabia.

If Saddam Hussein's regime is destroyed in a war, Iraq will be occupied by UN allies, which will include forces and administrators from Egypt, Turkey, Syria, and Saudi Arabia. The peace settlement can make certain that little opportunity is given either to Syria or to Iran to take over the destabilizing military superiority that Saddam enjoyed. Certainly the power of both Iran and Syria will be somewhat enhanced, at the expense of Iraq, but this will be a move back toward balance, not toward greater instability.

Negotiating a Peaceful Compromise

But suppose Saddam Hussein offers to negotiate, say, shortly before the 15 January deadline, between the 12th and the 15th. Will not that be a demonstration that the sanctions are accomplishing their purpose? Unquestionably, the President will be under great pressure to negotiate, from strong voices at home, and among our allies. And what do we negotiate? If Saddam withdraws from Kuwait, or from most of Kuwait, and if we hesitate, we are likely to be forced by world opinion, grateful for this demonstration of Iraqi reasonableness, to abandon the sanctions. Our allies (forgetting the perfidious aggression which triggered the crisis) will no longer be willing to fight. And we, also, probably will no longer have the stomach for it. And so, despite a little loss of face, Saddam will have won. In fact, he may even gain, rather than lose, prestige because he can assert—and rightly—that he faced down the most powerful nation on earth.

For many people that will not matter. They believe that a negotiated peace, whatever the terms, is preferable to a single American battle death. That is true, of course, if a shameful peace does not make a future war more likely. In this case, we know enough about Saddam to recognize that we are merely postponing war to a time when we will be in a less favorable position to prosecute it. This is the answer to Senator Nunn's much quoted observation—or was it a question?—that the use of force against Iraq is justified, but he is not sure if it is wise.

Then what should we do if such an offer to negotiate comes suddenly from Baghdad?

Unless we are prepared to abandon completely the principles and policies that led to our horribly expensive deployment of forces to the Gulf, we have no choice. We reject the offer. We demand immediate acceptance of *all* the resolutions of the UN Security Council since the Iraqi invasion of Kuwait (including the demand for reparations). And we place our forces on highly visible pre-hostilities alert. To provide some time for the significance of this to be recognized in Baghdad, we might extend the ultimatum's deadline for 48 hours, but no more. If these terms are not accepted by the expiration of the deadline—and the acceptance accompanied by the beginning of unambiguous physical withdrawal of Iraqi forces from Kuwait—we initiate one or more of the courses of military action discussed later in this book.

An early war is likely, but not inevitable. But a later, and worse, war *is* inevitable unless we are fully prepared to fight that earlier war now.

CHAPTER 3

The Thorny Problem of Command

The United Nations (UN) forces that are assembled in Saudi Arabia, the Persian Gulf, and neighboring land and sea areas of the Middle East, are made up of army, naval, and air force contingents from more than 40 nations. Between one-half and three-fourths of the troops, ships, and airplanes in these forces are from the United States. But the sovereignty of each of the nations from which the contingents come is absolute.

All alliances, throughout history, have had to face up to the problems of coordinating the military activities of contingents of different sovereign allies. But it is doubtful if any alliance, facing either the actuality or probability of war, has had to reconcile the sovereignty of so many separate and independent allies. This has been a matter of great concern to all commanders in the field, as well as to their senior military command establishments in their home capitals.

What, then, is the command relationship among the several contingents deployed in the Gulf region?

The Basic Issues

There are six principal facts, or issues, affecting this question of command.

In the first place, United States forces constitute roughly three-fourths of the military forces deployed in the Gulf region against Iraq. Even taking into consideration Turkish forces along the northern border of Iraq, and Syrian forces along Iraq's northwestern frontier, US forces make up approximately half of the total military deployment against Iraq. The United States Government is unlikely to place its forces under the command of any military leader from another country in the alliance, if only because no other country can match our commitment in men and treasure.

Secondly, most of the UN deployment is in the sovereign territory of Saudi Arabia. The non-Saudi forces are there by invitation of the Saudi government. That government would be most reluctant to permit an allied commander from another country to make military decisions affecting Saudi troops on their own national territory, or involving offensive operations based on Saudi territory that have not been approved in advance by the Saudi Arabian government.

Third, commanders of other national contingents now in and around Saudi Arabia represent proud sovereign governments which, under the best of circumstances, may be reluctant to entrust the security and integrity of their forces to a commander of another nationality. Under the existing (not the best) circumstances, these governments may have political sensitivities which make such command arrangements even more difficult. (For instance, feeling as he does about the relationship of the United States and Israel, and having made his people well aware of this feeling, how easy will it be for President Assad of Syria to put his troops under American command?)

Fourth, it is an undeniable truism, confirmed by innumerable historical examples, that lack of centralization among allied forces is not only inefficient, it is dangerous. One of the nine Principles of War—and the one most firmly endorsed by Napoleon—is "Unity of Command."

Fifth, while unity of command is essential, integration of national forces for combat purposes is extremely difficult, and in the situation in the Gulf is virtually impossible from a practical standpoint, over and above the political difficulties noted in issues 1, 2, and 3, above. This is the problem which NATO calls "interoperability," and which has not been fully solved even in NATO, after 41 years of alliance. *Any plan that involves the integration of forces of different nationalities simply will not work.* It will be possible, as was done in several instances in the Korean War, to attach relatively small units of one nation to the divisions or corps of another nation, and some sort of intra-formation coordination (*not* integration) can be worked out on the basis of a clear-cut command relationship. Such coordination will work best among forces that have had NATO experience, such as Americans, British, and French.

Sixth, there will be serious problems in assuring coordination of forces of different services (Army, Navy, Air Force, Marine Corps) even of a single nation in "joint" operations, as any American military man knows. Those problems will be compounded when the coordination involves forces of different nationalities as well as different services in "combined" operations. But some form of coordination is essential.

There have been a number of newspaper articles showing that these issues are very much on the minds of the UN allies in the Gulf, and suggesting that the issues have not been resolved. Furthermore, if they have been resolved, we are not likely to know about it until or unless combat operations begin.

Assumptions

In order to analyze this situation in a logical fashion, it has been necessary to formulate some assumptions about command. The following are the assumptions on which all subsequent analysis in this book are based.

1. There is no single allied commander or headquarters.
2. In military activities before the outbreak of war, and in combat activities based on locations in Saudi Arabia, the sovereignty of Saudi Arabia will be respected, without loss of integrity or modification of doctrinal procedures of the several national contingents.
3. The military pre-eminence of the United States in the alliance has undoubtedly been recognized formally or informally by the separate national contingent commanders, with some kind of commitment by these commanders to conform to US plans and operations.
4. National contingents have been, or will be, so deployed as to require a minimum of integration of procedures for effective offensive operational performance. Flowing from this assumption are the assumed details of deployments and command structure summarized below.

Assumed Deployments and Command Structure

a. UN forces in Saudi Arabia are organized as an army group of two armies (Eastern Army and Western Army) commanded by a UN Army Group Commander-in-Chief.
b. The Eastern Army is made up entirely of US forces, deployed in northeastern Saudi Arabia generally from the Persian Gulf in the east, to a point about 180 kilometers inland, due south of the intersection of the Iraqi, Saudi, and Kuwaiti frontiers. The western bound-

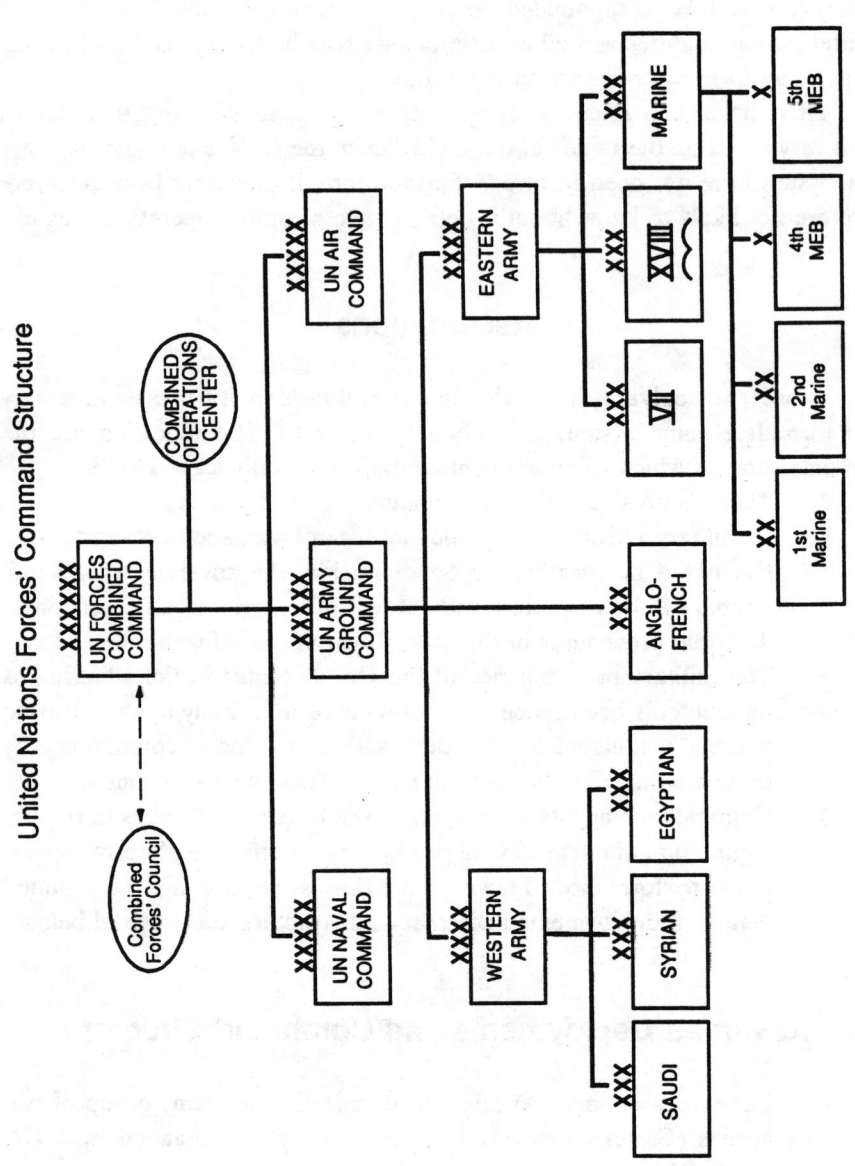

FIGURE 5. UN COMMAND STRUCTURE

ary of this US (or Eastern Army) Sector is a north-south line. (See map, Figure 7.)

c. British and French forces, in a provisional corps, are deployed in reserve behind the boundary between Eastern and Western armies. It is army group reserve, under the direct command of the UN Army Group Commander-in-Chief.

d. Egyptian forces are deployed in an Egyptian Sector, extending 60 kilometers to the west of the US Sector, which it adjoins on its right (east). The western boundary is a north-south line.

e. Other Arab forces (not including Saudi or Kuwaiti) are combined in a provisional corps under Syrian command. The Syrian Sector extends 60 kilometers to the west of the Egyptian Sector, which it adjoins on its right (east); the western boundary is a north-south line.

f. Part of the Saudi Arabian Army is deployed in a Saudi Sector, west of the Syrian Sector; there is no western boundary to this Saudi Arabian Sector.

g. The remainder of the Saudi Army, including the Kuwaiti contingent, is deployed in reserve behind the boundary between the Egyptian and Syrian Sectors. It is the reserve of the Western Army.

h. Saudi, Egyptian, Syrian, Kuwaiti, and other Arab forces are under the operational control of the Western Army, commanded by a Saudi Arabian general.

i. Overall command of operations of UN land forces is vested in a UN Army Group Command headquarters. For operations south of the Kuwaiti and Iraqi borders, the Western Army commander is the UN Army Group Commander-in-Chief; his deputy is the Eastern Army Commander. For operations north of the border, the Eastern Army commander is the UN Army Group Commander-in-Chief, with the Western Army commander his deputy.

j. The UN Combined Forces Commander-in-Chief is also the Commander-in-Chief of the US Central Command (General H. Norman Schwarzkopf). The UN Army Group comes under his command only in the event of war, and for operations north of the border of Saudi Arabia with Kuwait and Iraq. Also under his command are the UN Air Commander-in-Chief, and the UN Naval Commander-in-Chief. He is also Chairman of the Combined Forces Council.

k. Located at the UN Combined Forces Headquarters is a Combined Forces Council, consisting of the joint commanders of each nation contributing forces to Operation "Desert Shield." This Council is chaired by the UN Combined Forces Commander-in-Chief. The function of the Council is to coordinate operational plans, to reach agreement on the roles of each national contingent in those plans, and to agree on such other coordination measures and activities as may be required in activities involving so many national participants.

l. Within the UN Combined Forces Headquarters is a Combined Operations Center, to which is assigned at least one service member from each national contingent participating in "Desert Shield." The function of this Combined Operations Center is to work out the details of the participation of each national contingent in carrying out operational plans under the direction of the respective service commanders-in-chief (land, sea, air), to include unit assignments and missions, and to assure that coordination is so complete as to preclude interference by any one contingent with the performance of other contingents.

m. The UN Air Commander-in-Chief is also the commander-in-chief of US Air Forces under US Central Command. He is responsible for employing all UN air forces (other than naval air), which are either under his command or his direction, in a fashion consistent with the orders of the Combined Forces Commander-in-Chief and the guidance provided by the Combined Operations Center.

n. The UN Naval Commander-in-Chief is also the commander-in-chief of US naval forces under US Central Command. He is responsible for employing all UN naval forces (including naval air forces), which are either under his command or his direction, in a fashion consistent with the orders of the Combined Forces Commander-in-Chief and the guidance provided by the Combined Operations Center.

Perspective

In addition to the general and specific assumptions laid out in the preceding paragraphs, the reader can assume that—unless there is some indication to the contrary—the point of view expressed in the remainder of this book will be that of the UN Combined Forces Commander-in-Chief (or, the United States forces commander-in-chief, General Schwarzkopf). It is probably necessary to add that neither General Schwarzkopf nor anyone on his staff has been consulted in the preparation of this book, and that the authors recognize that they have neither his experience, his personality, nor the resources available to him. So he must in no way be considered responsible for the thoughts and decisions that are so glibly made in his name—explicitly or implicitly—in this book.

CHAPTER 4

How to Fight: The Operational/ Tactical Options

A military commander about to enter battle has two different sets of options: first, will he attack, or will he defend? Second, having made that choice, how will he fight?

Obviously, if we go to war in the Gulf, we shall already have made the first choice; we shall attack. So, in this chapter we shall address the various ways in which the forces of the United States, along with those of our UN allies, can attack the Iraqi forces of Saddam Hussein. These ways of fighting will relate either to operations, or tactics, or both.

The Operational/Tactical Issue

Non-military readers are often confused by the way in which military men use the terms "strategy," "operations," and "tactics." (Not to worry; many military men are often just as confused by the differences and relationships of these words.) Some explanation may help readers to avoid such confusion.

There are, of course, several kinds of strategy. For instance national strategy relates to the way in which national objectives are achieved and national policies are carried out. Contributing to national strategy are economic strategy, political strategy, and—not least—military strategy. Our concern in this book is military strategy, which governs the manner in which military forces are used to achieve national objectives, support national policy, and contribute to national strategy.

Military strategy is primarily concerned with the conduct of war at the highest level, and can encompass several theaters of war (also called theaters of operation). Strategy is "war on the map." Tactics, on the other hand, is the employment of troops on the battlefield, or "war on the ground."

In the past all warlike activity was conceptually considered to be either strategy or tactics. But in recent years the US Army has followed the lead of some other countries by dividing warfare conceptually into three elements, with "operations" inserted between strategy and tactics. The principal reason for this was to differentiate conceptually in a major, multi-front war between strategic thinking involving all or several theaters, and strategic thinking within a theater. So today the US Army uses the terms "operations" to refer to what might previously have been called "intra-theater strategy." For that reason, many military theorists think of operations as low-level strategy. Nevertheless, the Army considers operations to be a separate conceptual form of war, midway between strategy and tactics, and we shall endeavor to use the term in that fashion.

Thus, matters relating to the conduct of war in the Middle East/Gulf theater as a whole would be considered as "operational," or pertaining to "operations." Matters relating to combat at the corps, or division, or lower levels are considered to be tactical.

The Estimate of the Situation

When a military commander is contemplating a potential combat operation—offensive or defensive—he usually analyzes the various options that are available to him by means of what is known in the United States Army as an "Estimate of the Situation." In the British Army this analytic procedure is called an "Appreciation." All armies train their commanders and staffs to use some kind of systematic analytical method in preparing for combat. The classic, relatively simple, format used in the United States Army is shown in Figure 6.

Let us then use this format to get some idea of the kinds of things that are being considered by the commanders of the major United Nations forces in the Gulf region in their operational Estimates of the Situation.

> 1. Mission
> 2. Situation
>   a. Relevant Circumstances (Weather, Terrain, etc.)
>   b. Enemy Situation
>   c. Own Situation
>   d. Enemy Capabilities
>   e. Own Courses of Action (OCA)
> 3. Analyze Opposing Courses of Action
>   a. Enemy's Capabilities Affecting All OCA
>   b. Interaction of OCA with All Enemy Capabilities
> 4. Compare OCA
> 5. Decision (Who, What, When, Where, How, Why)

FIGURE 6. ESTIMATE OF THE SITUATION

I. The Mission

We can assume that the mission assigned to the UN Combined Forces Commander-in-Chief is:

A. To eject Iraqi forces from Kuwait, and facilitate the return of the legitimate government of Kuwait.

B. In cooperation with other UN forces, to secure Kuwait from any possible future Iraqi aggression.

The first portion of that mission is clear and unequivocal.

The second portion is deliberately vague and open-ended. This means the destruction of the resistance capability of opposing Iraqi forces, or their ejection not only from Kuwait but also from regions in southeastern Iraq adjacent to Kuwait. The UN Commander-in-Chief can expect adequate elaborating instructions from the UN Security Council or from the US Government. However, these instructions are likely to change as the situation develops.

II. Situation

A. Relevant Circumstances

The first of the relevant circumstances would be the specifics of the UN resolutions with respect to Kuwait, and particularly Resolution 678 which, in effect, specifies that offensive operations will not begin before 15 January.

The next relevant circumstance is the weather. The nature of the weather in the theater of operations is described in some detail in Appendix B of this book. Two principal considerations are particularly relevant:
1. Ground operations are difficult after about 20 March, because of the weather, and conditions for ground operations will not become reasonably suitable again until about 1 October.
2. Air operations can take place throughout most of the year, however they can be delayed, or their effectiveness degraded, by a variety of weather conditions which occur throughout the year.

The next relevant circumstance is the terrain. As will be seen in Appendix B, aside from a few sand dune regions, there are no severe terrain conditions affecting operations in most of the area.

But there is more than that to say about the terrain. This is country that favors the side with superior mobility, and a doctrine which emphasizes mobility. It is, in fact, ideal country for blitzkrieg-type operations, as Rommel so clearly demonstrated in the Western Desert in 1941 and 1942, and as the Israelis demonstrated in 1956, 1967, and 1973. And it is terrain where there is no need to take time to clear fields of fire, whether by saw, axe, explosives, or herbicides. This, again, favors the side which has, and uses, mobility, since this automatically offsets some of the defensive side's advantage in employment of firepower. Thus, once our troops are acclimated to the unfamiliar climate and temperature conditions, it is clear that the terrain is much more favorable to US forces than it is to the Iraqis.

One final relevant circumstance to be considered is the timing of the next significant Moslem holy season. This is Ramadan, which this year falls between 19 March and about 17 April. The principal significance of this holy season to this military analysis is that some of the UN Arab allies might be reluctant to undertake operations during Ramadan. However, this is not necessarily a limitation. Two of our allies—Egypt and Syria—initiated a major war against Israel during Ramadan in 1973. Furthermore, bad weather will be closing in on the region about 15 March, which probably renders the question moot.

B. Enemy Situation

A comprehensive survey of Iraqi forces is contained in Section II of Appendix D.

We are concerned about the deployment of Iraqi troops which could affect the mission of liberating Kuwait. Iraqi deployments are shown on the map of Figure 7.

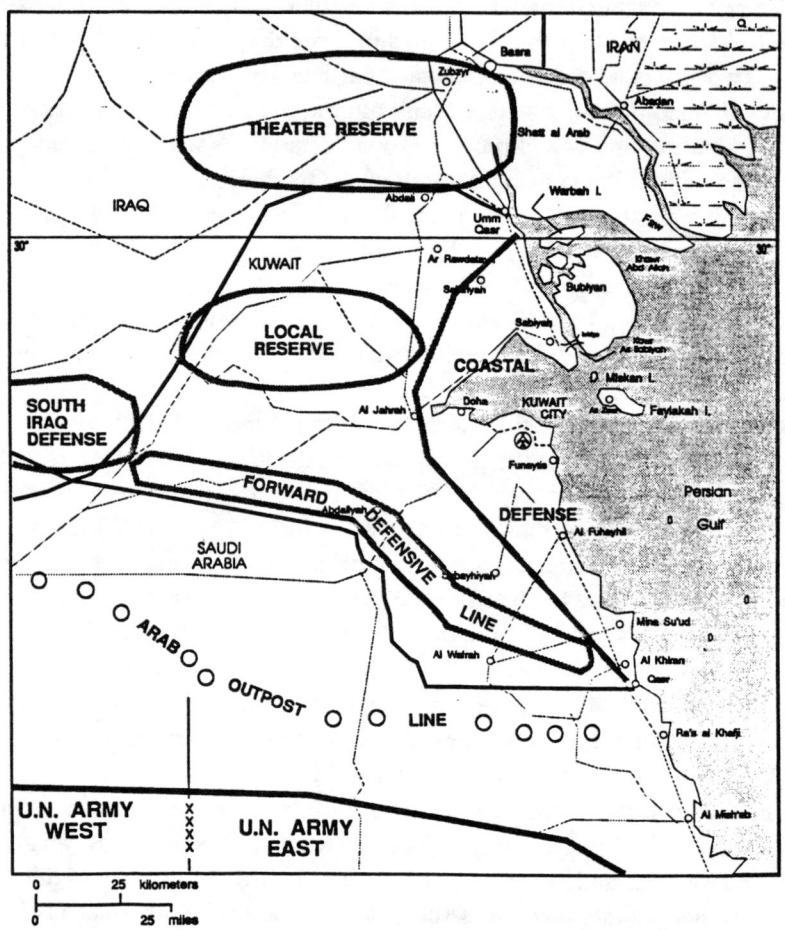

FIGURE 7. DEPLOYMENT SITUATION, KUWAIT AND VICINITY

By 15 January there are expected to be about 480,000 Iraqi ground troops in the region, organized in five corps of 28 divisions, approximately as follows:

1. Deployment

How to Fight: The Operational/Tactical Options

    a. A corps of about six divisions in defensive posture along the southern border of Kuwait with Saudi Arabia; about 84,000 combat troops. This Southern Defensive Force is heavily dug-in in a fortified zone, probably about 20 kilometers deep, running the width of Kuwait, north of the Saudi Arabian border.

    b. A corps of about six divisions deployed along the east coast of Kuwait; about 84,000 combat troops. This Coastal Defensive Force occupies a belt of field fortifications, 10 to 20 kilometers deep, running along the entire coast of Kuwait, from Bubiyan Island south to the Saudi Arabian border.

    c. A corps about four divisions deployed in central Kuwait, as a local reserve for both the Southern and Coastal Defensive forces; about 56,000 combat troops. This Local Reserve Force has extensive trenches and field fortifications, primarily intended to protect troops from air attack.

    d. A corps of about six divisions deployed in southern Iraq west of Kuwait, extending westward about 200 kilometers. There are about 84,000 combat troops in this South Iraq Defensive Force. Trenches and field fortifications are designed primarily to protect troops from air attack.

    e. The Republican Guards Corps of six divisions, deployed as a strategic reserve between Basra and northern Kuwait; about 90,000 combat troops. This Strategic Reserve Force has ample trenches and other field fortifications designed primarily to protect the troops from air attack.

    f. Miscellaneous service and support troops scattered throughout the region, generally west of Basra and south of An Nasariyah; approximately 80,000 troops.

2. The quality of these troops ranges from: "Above Average Regional Quality," to "Poor." The Republican Guards can be assessed as "Above Average Regional Quality." Most of the remainder are assessed as "Below Average Regional Quality." All Iraqi troops exhibited substantial tenacity in defensive posture during the Iran-Iraq War. However, Iraqi troops have not faced a first-rate opponent since 1973, when they performed abysmally against the Israelis.

Probably three-fourths of the Iraqi Air Force is based in central or southern Iraq, and available for operations in Kuwait and vicinity. The quality of the Iraqi Air Force is assessed as only "fair," in comparison to the high quality of the much more numerous air units of the UN alliance.

NUMERICAL COMPARISON OF IRAQI & UN FORCES

Category	Iraqi Forces	UN Forces US	UK/Fr	Other	Total	Ratio UN/Ir
Personnel	480,000	310,000	25,000	100,000	435,000	0.9/1.0
Divisions	28	10	2	6	18	0.6/1.0
Tanks	4,000	2,250	200	1,000	3,450	0.9/1.0
Combat Helos	120	500	50	100	650	5.4/1.0
Combat Aircraft	500	1,430	300	200	1,930	3.9/1.0

FIGURE 8. NUMERICAL FORCE COMPARISON

C. Own Situation

A comprehensive survey of the forces of the UN Alliance is contained in Section I of Appendix C.

1. US Forces
    a. Land forces. By 15 January the United States will have approximately eight Army divisions in two corps (XVIII Airborne Corps, and VII Corps) available for operations in the Kuwait area. In addition, there will be a corps of two Marine Corps divisions, the Eastern Army's reserve, plus two Marine Expeditionary Brigades (MEBs) afloat on the Persian Gulf.

    It is conservatively estimated that these 10 divisions are the equivalent, in terms of technology and firepower, of 15 Iraqi divisions. This estimate is based upon many characteristics of US weapons unmatched by the Iraqis. Among these are the superior firepower and maneuverability of the M1 (Abrams) tank; the awesome antitank and antipersonnel capabilities of our attack helicopters, particularly when operating with air superiority; the contribution which our transport helicopters make to the already substantial maneuver and economy of force capability of US ground forces.

    b. US Air forces. There are approximately 1,430 combat fixed-wing aircraft in units from the US Air Force, the US Navy, and the US Marine Corps. (Theoretically about another 150 naval combat aircraft could be available, but it is understood that the US Navy intends to rotate its carriers, keeping three in operational locations, and three further back for maintenance, replenishment, and recuperation.)

As with ground forces, the superiority of technology in general, and firepower in particular, can be represented by a factor of about 1.5. In other words, in terms of technical and material quality, the 1250 US aircraft are roughly the equivalent of 1875 Iraqi aircraft.

c. US Navy. In addition to the naval air contribution from three carriers (included in the air forces listed above), the Navy has a surface-to-surface firepower capability from approximately 40 warships in and around the Persian Gulf, Red Sea, Arabian Sea, and Eastern Mediterranean. Most of these vessels are in the Persian Gulf, where they add significantly to the technological ground force superiority noted above. Long-range gun and missile fire is available not only to support land forces across about one-third of the area of Kuwait, it is particularly important in providing firepower support to amphibious operations.

NUMERICAL COMPARISON OF IRAQI & UN FORCES CONSIDERING TECHNOLOGY

Category	Iraqi Forces	US	UK/Fr.	Other	Total	Ratio UN/Ir
Personnel	480,000	465,000	37,500	100,000	602,500	1.3/1.0
Divisions	28	15	3	6	24	0.9/1.0
Tanks	4,000	3,375	300	1,000	4,675	1.2/1.0
Combat Helos	120	750	75	100	925	7.7/1.0
Combat Aircraft	500	2,145	450	200	2,795	5.6/1.0

FIGURE 9. TECHNOLOGICAL FORCE COMPARISON

2. British and French Forces
   a. Land forces. These comprise two relatively small divisions. The technology and firepower of British and French forces are approximately comparable in quality to those of American divisions. Thus, in materiel terms, the British and French contingent is the equivalent of at least two full-size Iraqi divisions.

  b. Air forces. The British and French are providing approximately 300 combat aircraft. For similar reasons, these aircraft can be considered the material equivalent of about 450 Iraqi combat aircraft.
  c. Naval forces. The British and French naval contingents each have seven warships, one-third of approximately 40 warships contributed by UN allies other than the United States. These add to the already significant naval firepower available to support the land battle.
3. Other UN Allies
  a. Ground forces. These consist of two Saudi Arabian division equivalents, two Egyptian divisions, one Syrian division, and miscellaneous small contingents. This is a total Arab force of about six division equivalents, which can be considered to be approximately the technical and firepower equivalent of the same number of Iraqi divisions.
  b. Air forces. About 200 combat aircraft are contributed by the other UN allies. These are probably, on the average, somewhat more advanced in terms of technology and firepower than those of the Iraqi Air Force.
  c. Navy. More than 20 small warships are available to augment the naval support firepower of the United States.
4. Relative Combat Effectiveness

  In addition to the materiel qualitative superiority of US and some allied weaponry over the Iraqi weapons and equipment, there is a less easily-defined, but nonetheless substantial troop quality, superiority of the United States forces and those of some of the other UN allies over the Iraqis. This is in terms of relative combat effectiveness values, or CEVs. As discussed in Appendix E, this superiority is estimated as about a factor of 2.0 in the comparison of US, British, and French forces with the Iraqis.

  It is dangerous to underestimate an enemy. On the other hand, it is foolishly unprofessional and costly to overestimate an enemy whose quality can be reasonably assessed on the basis of historical experience and contemporary objective intelligence estimates. We are not privy to current intelligence estimates of the Iraqi Army. We are aware, however, of evidence that tends to confirm the historical evidence discussed in Appendix E.

5. The Effects of Defensive Posture

  There is a time-honored "rule of thumb" that an attacker should have a 3-1 superiority over a defender in order to expect victory. There is little validity to that rule in general, and none at all in making a comparison of forces that may be involved in a war in the Gulf region after 15 January.

The so-called "rule of thumb" is predicated upon a comparison of forces of about equal quality, equipped with weapons of comparable effectiveness. In fact, in 20th Century warfare the attacker, on the average, has had a strength less than twice that of the defender. And, again on the average, the attacker has been successful more than 60 percent of the time. The only way to make a quantitative comparison of the combat power of two opposing forces is by quantifying all variable factors affecting the forces under the circumstances of the battle, allowing for the multiplying effects of defensive posture, of terrain, of weather, of surprise, and of the relative quality, or combat effectiveness, of the opponents. Such comparisons are possible, and when made, a combat power superiority of 1.5-1.0 is invariably decisive.

NUMERICAL COMPARISON OF IRAQI & UN FORCES CONSIDERING TECHNOLOGY & EFFECTIVENESS

Category	Iraqi Forces	UN Forces US	UK/Fr.	Other	Total	Ratio UN/Ir
Personnel	480,000	930,000	75,000	100,000	1,105,000	2.3/1.0
Divisions	28	30	6	6	42	1.5/1.0
Tanks	4,000	6,750	600	1,000	8,350	2.1/1.0
Combat Helos	120	1,500	150	100	1,175	14.6/1.0
Combat Aircraft	500	4,290	900	200	5,390	10.8/1.0

FIGURE 10. TECHNOLOGICAL AND EFFECTIVENESS FORCE COMPARISON

By 15 January the Iraqis on the southern front will have a significant numerical superiority over the UN allies. However, a combat power comparison of the UN allied forces and the Iraqis in Kuwait and vicinity, reveals an allied combat power superiority between 1.5 and 2.0 to 1.0, with an even higher value in favor of US forces making the main efforts in the various courses of action that will be considered in subsequent chapters.

This comparison presumes that the Iraqi forces survive the preliminary air strikes of UN air forces more or less intact. If they should be as badly damaged and demoralized by air attack as were the Egyptians, Jordanians, and Syrians in the 1967 Six-Day War, the combat power ratio will be even greater, and losses of the US ground forces, and of other UN allies, are likely to be far less than most of the "expert" 'estimates that have been appearing in the press.

Figures 8, 9, and 10 show how the quantitative or numerical comparison of Iraqi and UN forces is affected by the successive application of factors reflecting technology superiority, and combat effectiveness superiority (both of which favor the UN allies). Figure 11 shows the result of applying the factors reflecting the effects of posture and terrain, both of which favor the Iraqi defenders. In Figure 11 the posture factor considers that a portion of the Iraqi troops are in fortified defense, with a multiplying effect of 1.6, but that most of them will be in hasty defensive posture, with a multiplying effect of 1.3, yielding an average posture factor of about 1.4. There is a modest 1.1 terrain factor in favor of the Iraqis. The resulting 1.54 factor is applied to the numbers of Iraqi troops and of tanks; it will have no effect upon helicopters or fixed-wing aircraft.

NUMERICAL COMPARISON OF IRAQI & UN FORCES CONSIDERING TECHNOLOGY, EFFECTIVENESS, & POSTURE

Category	Iraqi Forces	UN Forces US	UK/Fr.	Other	Total	Ratio UN/Ir
Personnel	739,200	930,000	75,000	100,000	1,105,000	1.5/1.0
Divisions	43.1	30	6	6	42	1.0/1.0
Tanks	6,160	6,750	600	1,000	8,350	1.4/1.0
Combat Helos	120	1,150	150	100	1,750	14.6/1.0
Combat Aircraft	500	4,290	900	200	5,390	10.8/1.0

FIGURE 11. TECHNOLOGICAL, EFFECTIVENESS, AND POSTURE FORCE COMPARISON

6. Logistics Considerations

There must be some question about the complete logistical readiness of US forces (particularly Army forces) to conduct an intensive, sustained campaign by 15 January. On the other hand, there is much more serious question about the logistical capability of the Iraqis to conduct such a campaign, particularly since anticipated US and allied air supremacy will almost assuredly result in severe damage to Iraqi military depots and stockpiles, as well as in the destruction of Iraqi supply lines. (Logistical considerations are considered in detail in Chapter 10.)

7. Special Identification Considerations

The problem of "identification, friend or foe" (IFF), which is behind the incidence of "fratricide" in modern, high-speed, high-technology warfare, will be at least doubly exacerbated in the coming conflict. In the first place the Iraqis have much of the same ground force and air force equipment as some of our allies, which can lead to misidentification and mistaken engagement of friendly forces. This may slightly favor the Iraqis. In the second place, misidentification of targets of normally easily-distinguishable characteristics will be compounded by the swirling dust of the desert floor.

D. Enemy Capabilities

There are two general Iraqi capabilities which cannot be ignored.

In the first place, it is believed that the Iraqis have about 36 Soviet SCUD-B missiles, with a range of more than 300 miles, enabling them to hit Riyadh, the Saudi Arabian capital, and many of the major Saudi Arabian oil fields, to say nothing of the UN forces and command posts north of Riyadh. These missiles can also easily hit Jerusalem and Tel Aviv in Israel. They are, however, very inaccurate. Nevertheless, the employment of these missiles in a pre-emptive attack against targets in Saudi Arabia could be damaging. If used against Israel, retaliatory action by the Israeli Air Force could place great strains on the UN alliance, since the Arab allies will not want to be co-belligerents with Israel against another Arab country.

Secondly, the Iraqis could use chemical and biological weapons at any time, either as integral elements of defense plans, or in a surprise move during the hostilities. This capability is, of course, unquestionable. Iraq used such weapons in its recent war with Iran, and has used them in brutal suppression of internal opposition by Kurdish separatists. As in World War II, however, the use of these weapons against an opponent with comparable retaliatory capability is unlikely. Chemical and biological weapons are inherently incapable of producing a quick or decisive victory. Any short-term, local advantages that might be obtained by their use will be offset by the delays inherent in undertaking protective measures not only from the effects of their own weapons, but also from retaliation by UN forces.

The principal Iraqi capabilities which could be employed against the UN forces, and particularly against US forces in the Eastern Army, are listed briefly below:

1. There is the possibility of a pre-emptive attack against the Eastern Army on the Kuwait front.
2. There is also the possibility of a pre-emptive attack against the UN allies of the Western Army.
3. The current Iraqi plan appears to be to defend in depth all along the defensive line they have established in southern Kuwait, and further west along the frontier with Saudi Arabia.
4. Another possibility would be to withdraw from all of Kuwait except for Bubiyan and Warbah Islands and the northeast corner of mainland Kuwait in the Umm Qasr area; a new defensive line would then be established to protect this portion of Kuwait and southeastern Iraq.

The next step in the Estimate of the Situation is for the commander and his staff to consider the various courses of action that might possibly be used to carry out the mission. Before doing that, however, it would be helpful to review conceptually the kinds of action that can be performed by an attacking military force.

Methods of Offensive Warfare

There are three different ways (or maneuver options) of waging offensive operational or tactical warfare: penetration, envelopment, and blockade or siege.

A frontal attack is an attempt to punch a hole through (or penetrate) an enemy's defensive line by direct assualt. It is also usually the most costly in terms of casualties to the attacker. From the standpoint of the defender, it is easier to meet and to block a frontal attack than any other sort of offensive maneuver.

An envelopment is an effort by an attacker to avoid the costs of a frontal attack by moving around the flanks of the defender's position to strike him on the flank, or in the rear. From time immemorial, enveloping maneuvers have tended to unnerve defenders, because they cannot protect their flanks, or rear, as readily or as confidently as they can their front.

There are two basic forms of envelopment. One of these, the simple, or "close-in" envelopment, is directed at the flank or flanks of the defender. (See Figure 12.) The other form is a "wide" envelopment (also called a "turning movement," or a "strategic" envelopment) which goes around and past the enemy's flank, either to threaten his line of communications to the rear, or to strike him from the rear, or both. (See Figure 12.)

A variant of the envelopment is the double-envelopment, which strikes at, or goes around, both the flanks of the defending force simultaneously. (See Figure 12.) This can be either two simple envelopments, or two turning movements or (rarely) one of each. It is also rare for an attacking commander to attempt a double envelopment unless he has a great strength preponderance over the defender.

Sometimes the strength of the defender's position is such that a frontal attack, to achieve a penetration, would be impossible, or so costly in casualties as not to be worth considering. And sometimes the attacker cannot find an exposed flank to envelop, and thus has no choice but a frontal attack to make a penetration. This could be either because the defensive line stretches between two barriers that are impassable to the attacker, or because the defender has all-around protection, and thus no exposed flank. In such a case, if the attacker is not willing to incur the cost of a head-on frontal attack, he can resort to a blockade, or a siege. In this event, the attacker deploys his forces along the front of the defender, but does not mount a major attack, although he may resort to raids or probes against various portions of the defender's position.

The term blockade usually refers to military activities along an extended front or fronts, and can be either a land blockade, or a naval blockade, or both. A siege, on the other hand, refers to a geographically more limited operation against a relatively small defended area, usually a city or a fortification. In a siege, the attacking, or besieging, force usually encircles the defender's position. A blockade may or may not imply encirclement.

To recapitulate, an attacker has the choice of penetration, envelopment, or siege-blockade, and usually will adopt some combination of two or all three of these options.

II. Situation (cont.)

E. Own Courses of Action

Bearing in mind the traditional offensive maneuvers discussed above, on the map, noting the deployments of the opposing forces, and considering the capabilities of the forces of the Eastern Army, four possible courses of action come quickly to mind:

How to Fight: The Operational/Tactical Options

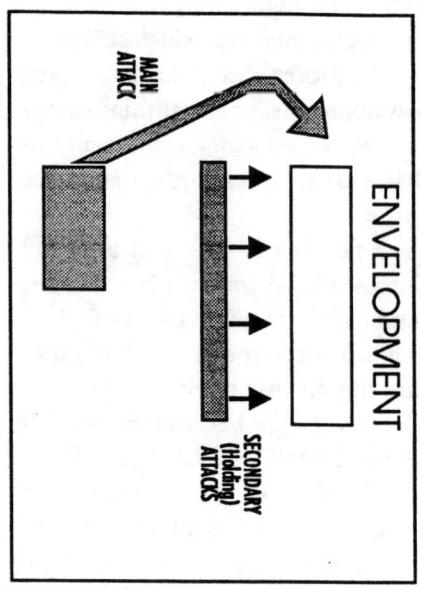
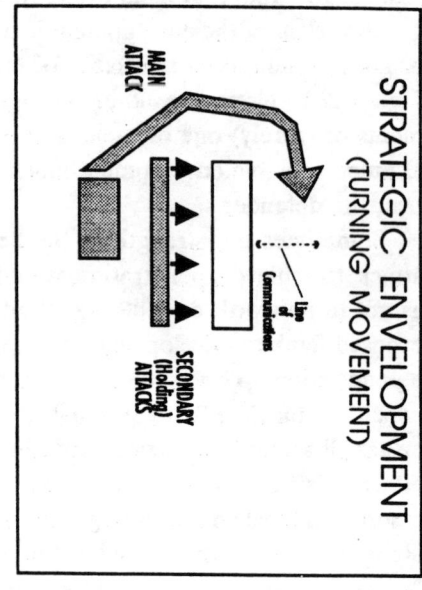
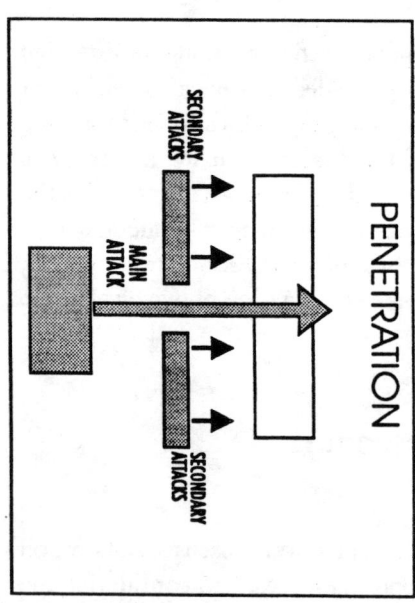
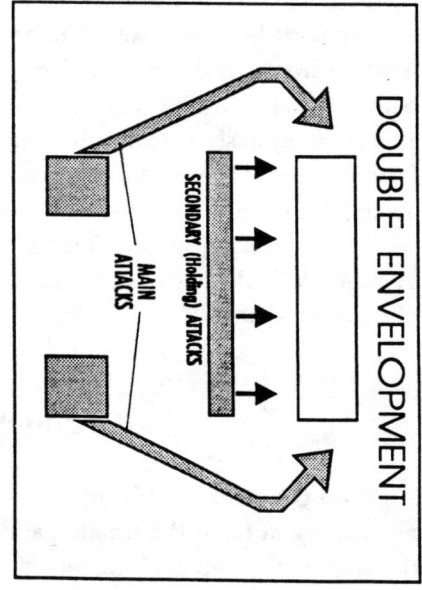

Figure 12. Basic Combat Maneuvers

First, a direct, frontal blow at the principal Iraqi defensive line in southern Kuwait to achieve a quick and decisive solution;

Second, a double envelopment of that defensive line, using mobile armored forces for the left hook, and the amphibious capability of the US Marine Corps, supported by the US Navy, to carry out the right hook by coming ashore from the Persian Gulf;

Third, a wide, "strategic" clean up envelopment or turning movement, with mobile armored forces coming into west, central Kuwait from the desert behind the Iraqi defensive line in southern Kuwait, possibly accompanied by a deeper Marine amphibious operation north of Kuwait city, for either a single or double strategic envelopment; and

Fourth, an active blockade, or siege, of the Iraqi forces, not only in Kuwait but in all of Iraq, highlighted by limited attacks and probes, forcing the defenders to use up valuable supplies and spare parts (in short supply because of the embargo). When a weak spot is found, it would be exploited by a limited or major offensive. If possible, this action on the southern front should be matched by similar moves by Turkish forces in the north, and Syrian forces in the northwest.

There is one problem with these four courses of action. None of them appears to exploit two areas of significant superiority of the UN forces—particularly the superiority of the US forces of the Eastern Army. These two areas are, first, the tremendous American and allied superiority in air power (fixed wing and rotary wing), and second the substantial air mobility, and air assault capability, of the US ground forces.

Bearing in mind the airpower, and air assault capabilities, of the US forces, another review of the possible options for courses of action is required. This gives us five likely courses of action, none of which is mutually exclusive.

Operation "Colorado Springs"

First is an intensive, comprehensive air campaign, code-named, for this study, Operation "Colorado Springs." This would be carried out in two major phases. In the first phase the UN air forces would strike at military targets all over Iraq and Kuwait, focussing in particular on: air bases (and Iraqi planes in the air); air defense installations; national, regional, and field forces command centers; electronic communications centers; long-range missile launching sites; military supply stockpiles of all kinds, especially stockpiles of chemical and biological weapons; plants manufacturing chemical and biological weapons; potential nuclear weapons plants; other weapons plants; road and railroad lines of communications; troop concentrations and assembly areas in Iraq and Kuwait.

This phase of the air campaign could last a week or more, but within the first 48 hours it should result in completely clearing the Iraqi air force from the skies, and in the destruction of most of the Iraqi air defense capability. This would permit the initiation of the second phase of "Colorado Springs," while the first phase would still be continuing.

In the second phase of the air campaign, UN air forces would concentrate against Iraqi ground forces located in and near Kuwait, to include: any air defenses that may have survived Phase I; command posts at all levels; artillery positions; infantry defensive field fortifications. How effective "Colorado Springs" will be in damaging and destroying the military effectiveness of the Iraqi ground forces is debatable. It must be remembered that in the history of airpower to date, no air campaign has been able by itself to defeat or destroy hostile ground forces. And substantial elements of ground units, when well dug-in, have been able to survive the most intensive air attacks. On the other hand, the circumstances of this operation are different from past efforts of air forces to "go it alone." The terrain and vegetation (or lack thereof) favor air attacks to an exceptional degree. And the air forces will be operating with weapons of hitherto unsuspected accuracy and effectiveness against fortified targets. Given these new circumstances, and considering recent historical examples in the 1967 and 1973 Arab-Israeli Wars, the possibility that airpower alone can cause such devastation, destruction, and demoralization as to destroy totally the effectiveness of the Iraqi ground forces cannot be ignored.

Until, however, such a capability is demonstrated in practice, it must be assumed that the Iraqi defenders will remain effective at least until the UN ground forces get involved. And discussions of the subsequent courses of action will assume that they do retain an effective defensive capability despite the air attacks.

Operation "Bulldozer"

This course of action presumes that at some time after the initiation of Phase 2 of "Colorado Springs," a frontal assault will be made against the Iraqi defensive positions in Southern Kuwait, The object will be to overwhelm those defenses in general, and penetrate them at one or more specific spots. Facilitating this main frontal assault will be flanking attacks by mobile armored forces in the desert, and by amphibious forces from the sea. It is also assumed that the actual breakthrough will be accomplished with the assistance of an air assault at or near the rear boundary of the fortified zone, coordinated with a frontal, penetration thrust.

Operation "Leavenworth"

In general the Tactics Department at the US Army Command and General Staff College at Fort Leavenworth, Kansas, favors enveloping maneuvers over simpler, and usually costlier, frontal assaults. Thus the "approved solution" of a Leavenworth tactical problem will typically include the use of mobile forces for envelopment, combined and coordinated with an air assault. This course of action, a "Leavenworth" solution, envisages a close-in envelopment of the right flank of the Iraqi defensive line in southern Kuwait, coordinated with an air assault close behind that right flank, while a Marine amphibious assault on the right flank of the Iraqi defenses completes a double envelopment.

Operation "RazzleDazzle"

Leavenworth usually frowns on more risky and more complicated plans, believing firmly in Simplicity—a Principle of War—and in Murphy's Law. (If it can go wrong, it will go wrong.) On the other hand, operations that may be unduly complex for some military forces, can be taken in stride by well-trained professionals of the sort now deployed in the Gulf. Recognizing that complexity introduces great risks of failure, this course of action suggests that the risk is justifiable. It envisages a sweeping turning movement across the desert into western Kuwait (behind the bulk of the Southern Iraq Defense Force, presumably pinned down by Phase 2 of "Colorado Springs"). At the same time another turning movement will be conducted from the sea, with an amphibious assault across Bubiyan Island. Completing the encirclement will be a deep air assault northwest of Kuwait City. The individual elements of this triple envelopment will leave blocking forces to prevent interference by the Republican Guards reserves to the north, and will then drive south to encircle and destroy the remaining Iraqi forces in Kuwait.

Operation "Siege"

Good reasons can be offered (logistics, lack of readiness, command problems, strong residual Iraqi defensive capability) as to why the UN ground forces should not immediately initiate a major ground campaign. Rather, shortly after the initiation of Phase 2 of "Colorado Springs," the UN forces could initiate limited probes all along the Iraqi defensive lines in Kuwait and to the west, seeking for weak spots, while requiring the Iraqis to undertake considerable activity to counter these probes. If possible, there should be similar probes by Turkish force along Iraq's northern frontier, and by Syrian forces along the northwestern frontier. Then, after the situation is clarified, after Iraqi forces have been weakened by air strikes, by their own logistical problems, and by the confusion created in meeting the UN probes—the Eastern Army would initiate either "Bulldozer," or "Leavenworth," or "RazzleDazzle," to bring the conflict to a quick conclusion.

* * *

Each of these courses of action will be examined in more detail in subsequent chapters.

CHAPTER 5

Option One: Operation "Colorado Springs"

Ever since airpower first played a significant role in war—in World War I—airmen have been dreaming of the time when airpower, acting alone, could win a battle, or a campaign, or a war. That such a day would come seemed inevitable to the principal apostle of airpower, Giulio Douhet, and he expected it would be soon after he wrote his book *The Command of the Air*, which was published in 1921. In the 1920s and 1930s similar concepts of the ubiquity and coming omnipotence of airpower were stridently expressed by US Army Major General William Mitchell and, in Great Britain, by the father of the Royal Air Force, Air Marshal Sir Hugh Trenchard.

Disciples of these apostles of airpower expected that the dramatically improved effectiveness of combat aircraft in World War II would confirm the prophecies of the supremacy of airpower on land battlefields. But it did not happen. The one clear-cut effort by American and British airmen to defeat a ground army (Operation "Strangle," in Italy, in early 1944), severely punished the German forces in Italy, but did not bring them close to defeat. Subsequent operations did demonstrate, however, that when used in coordination with effective ground forces, airpower could make a major, often decisive, contribution to ground force success.

Undoubtedly the Allied strategic air campaign against Germany in World War II played a major role in the defeat of Hitler's Reich. But the Germans did not surrender, and continued to mount effective defenses, and even counteroffensives (as in the Battle of the Bulge), against both ground and air attack, until finally overwhelmed by vastly superior ground forces, aided by effective air support. This was also true in the war against Japan, where land-based and carrier-based airpower was a major factor in the defeat of Japan in the summer of 1945. But it was not the only factor, and no more significant than the naval blockade of Japan, combined with the defeat of Japanese armies and navies in the Pacific and in Asia by Allied land and sea forces.

Israeli airpower may have been the most important factor in Israel's overwhelming victory in the 1967 Six Day War. Its full significance in that war was not realized, however, even by the Israelis, until the next Arab-Israeli war, in October 1973. In that war effective Egyptian and Syrian air defenses (provided by the USSR), prevented the Israeli Air Force from dominating the skies. Israeli soldiers, who had become contemptuous of the Arab armies because of their poor performance in the 1967 war, now learned that Arab troops could fight tenaciously if they had protection from the hostile air force. In retrospect, it was clear that the sudden Arab collapses in 1967 had been largely due to the devastating effect of unopposed Israeli airpower. Even so, the Israeli ground forces did encounter some opposition on the ground, and were an essential element in the Israeli victory.

Interestingly, there had been a very similar demonstration of the devastatingly demoralizing effect of airpower against unprotected ground forces in that same arid, treeless part of the world nearly 50 years before the Six Day War. This was the effect of attacks by General Edmund Allenby's small air contingent upon retreating Turkish troops during the Battle of Megiddo, in September 1918. The Turks—who throughout the war up to that date had demonstrated remarkable toughness, determination, and stolidity under fire—panicked. That incident undoubtedly contributed to the Turkish request for an armistice less than a month later. But it was primarily the rapid pursuit by British horse cavalry, into Damascus, and then north to Aleppo, which caused the Ottoman Empire to acknowledge its final defeat.

These experiences of airpower, despite its undeniable value as an adjunct to ground power, have caused military theorists to conclude that airpower can never win a war by itself. Airmen, no matter what their innermost thoughts may be, are particularly emphatic in asserting that it is not the purpose of airpower to defeat a ground enemy. They emphasize—and quite properly—that this is accomplished by the interaction of air and ground power. Since land forces are undeniably essential in the occupation and holding of enemy terrain, it has been assumed that no enemy ground force would surrender until attacked on the ground, no matter how devastating the air attack may be.

Indeed, this may be true.

The situation in the Gulf region today, however, is so different from previous instances when airpower has been employed, that it just might be possible for airpower to win the coming war, without any need for extensive combat employment of ground forces. Although, of course, they would be essential for accepting surrender and occupying the hostile territory abandoned or surrendered. In any event, if hostilities break out between Iraq and the UN allies, there will be an opportunity for airmen again to try to win a war single-handed. Certainly their efforts to seize such an opportunity cannot in any way impair the overall UN conduct of hostilities against Iraq. We expect that such efforts will be made.

The Concept of Operation "Colorado Springs"

We have no way of knowing, of course, what the actual plans of US and other UN planners are or will be. We may be certain, however, that the UN forces will employ the tremendous airpower superiority available to them in some way similar to the concept suggested in this chapter. We have called our speculative air campaign Operation "Colorado Springs," after the location of the US Air Force Academy, exemplifying the young (as compared, for instance, with West Point) traditions of the US Air Force.

Phase One

We believe that "Colorado Springs" should be carried out in two phases. The first phase would be devoted to destroying the offensive and defensive air warfare capabilities of Iraq, while at the same time generally degrading, or destroying the ability of the nation to support the conduct of any other type of war activity in Kuwait, or elsewhere.

The targets of the first phase of Colorado Springs would be military and, in an approximate order of priority, would be the following:

- The Iraqi Air Force, at its bases and in the air;
- Air defense installations in general, particularly those defending air bases and subsequent targets;
- Command centers at all levels: national, regional, and field forces in Kuwait and vicinity;
- Electronic communications centers at all levels;
- Long-range missile launching sites;
- Military supply depots, and all identifiable stockpiles of military equipment;
- Stockpiles of chemical and biological weapons;
- Production facilities for chemical and biological weapons;

- Production facilities for potential nuclear weapons;
- Other weapons' production facilities;
- The road and railroad network, particularly sensitive points such as bridges.

Every effort will be made to achieve surprise in an initial massive attack, which would probably be directed mainly at air bases and major command headquarters and electronic communications centers. It will not be easy, of course, to achieve surprise; no other war in history has been so long advertised in advance. It is doubtful if the attack would start at midnight of 15-16 January, since that would be a time of maximum Iraqi alertness and preparation. Probably there will be a delay of several hours, perhaps as much as several days.

But there should be no warning as a result of significant, and unusual, ground activity before Phase One of the air assault begins. There will be no need for any such pre-assault ground movement, because there will be no major UN ground activity for at least two or three days after the air offensive begins.

Four sets of targets vie for first priority to be attacked. Obviously there can be no more important primary sub-objective than to destroy the Iraqi Air Force as quickly as possible. Wuch destruction would prevent Iraqi aircraft from interfering with, and delaying the progress of the air campaign, to assure its early effectiveness, and to protect American and allied airmen from possible losses at the hands of Iraqi planes.

It is no less important to destroy as soon as possible the effectiveness of the Iraqi politico-military leadership and e military high command, and their ability to communicate with subordinate headquarters. Hand in hand with both of these objectives will be the necessity for destroying the air defenses protecting the installations being attacked and the command and control centers from which Iraqi aircraft could be directed against the attackers.

Yet none of these vital targets is more important than the long-range Scud missiles which could be used to direct chemical—and particularly biological—warheads against UN forces and headquarters, and (given what we know of Saddam Hussein's thinking) against civilian targets in Israel.

Fortunately, allied air resources are sufficient to attack all four of these high-priority categories simultaneously. A variety of aircraft will be employed in the first day's massive effort against these four categories of targets. Probably more than 1,000 aircraft will take part in that effort, with an average of two sorties per plane, for a total of about 2,000 sorties. Probably as many as 20 or 30 aircraft—between 1 percent and 1.5 percent—will be shot down on the first day, most being lost to air defense missiles and AAA guns. Losses in subsequent days will decline, as more and more of the air defenses are knocked out. By the end of a week nearly 10,000 sorties will have been flown against Phase One targets, and the total loss will probably average less than 1 percent per sortie over that week, for a total of perhaps 60 to 70 aircraft. Once the main air defenses have been eliminated, losses will decline to two or three aircraft a day, out of around 1,000 sorties per day.

Bearing the brunt of the early Phase One effort would be F-111 fighter-bombers (many based in Turkey), and F-117A stealth fighters. Presumably Navy Tomahawk cruise missiles, from cruisers and battleships in the Persian Gulf and Red Sea, would also be used in these attacks. It is possible that B-52 bombers, probably staged through Diego Garcia in the Indian Ocean, would also take part in these attacks.

Should any Iraqi aircraft attempt to interfere with these strikes, they would be picked up immediately by the radars on Saudi and Air Force E3 AWACS and, if over water, Navy E-2C Hawkeyes, all of which would be keeping all of Iraq and its airspace under surveillance. Within minutes of takeoff these Iraqi aircraft would be engaged by loitering Air Force F-15 or F-16 fighters, or by Navy carrier based fighters—F-14s and F/A-18s—guided to their targets by US Air Force or Saudi AWACS or Hawkeyes.

Some percentage of Iraq's 600-odd combat aircraft would survive the first day's attacks, particularly those sitting in hardened shelters scattered around the air bases. But few of these would be able to take off from those bases in subsequent days because the runways would have been deliberately cratered by specially-designed concrete penetrating bombs. Efforts to repair the runways would be discouraged by randomly-timed explosions of delayed fuse anti-personnel bombs also scattered on and around the runways.

By the end of the second 24 hours the Iraqi Air Force would literally have been destroyed. Aside from continuing surveillance of all the wrecked air bases, the bulk of the attacking aircraft could then turn to other Phase One targets.

Meanwhile, the ability of the Iraqi high command to direct and control the armed forces would be steadily degraded. The national command headquarters and its related electronic communications centers in and near Baghdad would have been relentlessly pummelled. Most of the hardened underground shelters would almost certainly have survived, but ingress and egress to them would have been rendered difficult, and in some cases impossible. But the surface facilities would have been flattened, and communications between Baghdad and many outlying areas—including the forces assembled in southern Iraq and Kuwait—will have been seriously degraded.

Other command and communications centers throughout the country, and particularly those controlling forces in the south, and opposite the borders with Syria, Turkey, and Iran, will have been similarly pounded. Undoubtedly some surface and radio communications would still be operating, but the normal flow of messages would have been reduced to a trickle.

Also by the end of the first day, possibly within the first hour or two, the launching sites for the Iraqi Scud-B, long range missile will have been destroyed. If the attackers are able to achieve the surprise they would like and hope for, the crews of these launching sites will not have had time to get off any of their missiles. If surprise is not achieved, or if Saddam Hussein is able to preempt the UN attack, then several Scuds will probably have been fired before the sites are destroyed. Some of these would probably be fired in the direction of Tel Aviv and Jerusalem, with others directed at Riyadh, and known UN force command centers. Undoubtedly some of these missiles would claim some lives, but it is doubtful if they could have much effect because of the inaccuracy of these weapons. Fear of retaliation in kind will, it is believed, keep the Iraqis from using chemical or biological warheads.

During the first two days, UN air attacks would undoubtedly have been made against other Phase One targets. High in priority among these would be other launching facilities for chemical and biological weapons, and stockpiles of chemical and biological weapons.

Initially, F-111s and F-117s will be heavily engaged, particularly at night. Once the air superiority battle has been won, and most Iraqi air defenses destroyed or suppressed, they will be joined by other aircraft, to include: F-15s, F-14s, Tornado F.3s (from the RAF) and F-18s. Interdiction will be the responsibility of F-15Es, F-16s, and Tornado F-1s. More vulnerable aircraft—like B-52s, A-6s, A-10s, Jaguars, and Apaches—will join for selected targets after all Iraqi anti-air capability has been eliminated.

As requirements lessen for attacks on air bases, air defenses, and command facilities, increasing attention will be given to the Iraqi chemical and biological warfare capabilities, including production facilities. Nuclear reactors and potential nuclear weapons production sites will also be attacked. Higher priority will now be given to all kinds of military depots, particularly ammunition and weapons. Ammunition plants and weapons factories will also be attacked. More attention will also be given to roads and railroads, and to traffic that may be on them. Simply knocking out a few bridges will bring railroad movement to a standstill. Bridge-busting will also ruin the road network in the Tigris-Euphrates valley, forcing any remaining Iraqi traffic to rely increasingly on the desert roads, where it may be more easily discovered, attacked, and destroyed.

Option One: Operation "Colorado Springs"

By the end of the first week the assignment of aircraft to Phase One targets will be mostly dependent upon assessment of the damage already created, and any indications—by observation or by intelligence means—that any earlier attacks were less effective than had been anticipated.

Phase Two

By this time also, or perhaps earlier, a shift of much of the air attack capability to Phase Two targets will probably be made. These will be military targets in Kuwait and in southern Iraq. The following are the principal Phase Two targets:

- Regional command headquarters, to the extent these have survived Phase One;
- Air defenses among the operational ground forces from Basra and Kuwait City westward;
- Helicopter units;
- Artillery positions;
- Local depots, particularly ammunition;
- Any and all troop concentrations, with emphasis on armor;
- Field fortifications.

There is reason to believe that the nature of the Iran-Iraq War was such that the Iraqis had little concern about, or experience with, air attack and air surveillance. Thus, while they have ample numbers of air defense weapons, they seem to have neglected camouflage. Whether or not this is so, it is difficult to camouflage in the desert.

As shown above, the first priority in Phase Two will be the local air defense weapons, missiles and guns. Using sophisticated electronic means to find and destroy the radars, and then the weapons, the UN air attackers should deal with these as quickly and effectively as with the air defenses at airfields and other installations attacked in Phase One. This will be facilitated by specially equipped aircraft to pinpoint Iraqi electronic installations, to include EF-111, EA-6B and other "Wild Weasel" EW/ECM (electronic warfare/electronic countermeasers) planes.

Once most of the hostile air defense weapons have been found and destroyed, the UN aircraft will have a field day attacking the unprotected ground units. Accurate, "near zero CEP" (weapons that can strike within a few yards of selected targets) will be employed against hardened targets like bunkers, dug-in tanks, dug-in artillery pieces, and fortified command posts. Area bombardment from B-52s, flying so high they can be neither seen nor heard, will have a terrifying effect upon the Iraqi troops, no matter how well they may be dug in.

Augmenting these air weapons will be long-range missiles from Navy ships in the Persian Gulf, one-ton high-explosive projectiles from the 16" guns of two battleships, missiles from ground-based Army MLRS (long-range multiple rocket launchers), and Army and Marine Corps heavy, long-range artillery. Also contributing, with careful coordination, can be the Army and Marine Corps attack helicopters. This will probably be the most devastating anti-personnel bombardment in the history of warfare.

Also, it will be around-the-clock pounding. Air attacks will not be as precisely targeted at night as in daytime, but they can still be quite accurate. The Iraqis will have little or no capability to move supplies from the rear to the forward troops at night, because such movement will be detected and attacked.

There is, of course, serious question whether there will be any supplies to move forward, since supply depots will have been quite thoroughly smashed. And if supplies should remain in some places, will enough trucks survive to move them? Of course some supplies can be stockpiled in the forward areas, but these, also, will have been targeted by Air Force, Navy and Marine Corps aircraft, by Army and Marine Corps helicopters, and by long-range naval and land-based surface-to-surface missile and gun bombardment.

Since the combination of accuracy and massive firepower will be unprecedented, no one can be certain what the reaction of the surviving Iraqi soldiers will be. Military theorists, including ground soldiers, who did not believe that airpower could cause an army to surrender are now having second thoughts.

It is possible that individual units, major formations, or perhaps the entire Iraqi Army in Kuwait and vicinity will surrender rather than continue to endure this almost unimaginable punishment.

Operation "Salah-al-Din"

Should there be either a general surrender, or a mass surrender and collapse, of Iraqi forces in Kuwait, the Eastern Army will be ready to mount immediately a special operation—Operation Salah-al-Din—to liberate Kuwait City. The forces to do this will be the following:
- Task Force commander: Saudi Arabian major general
- Provisional UN armored brigade, commanded by a Kuwaiti brigadier, and including:

- Kuwaiti armored battalion (from Western Army)
- Saudi Arabian armored battalion (from Western Army)
- US armored battalion (from Eastern Army)
- US Marine Expeditionary Brigade (afloat)

Upon orders from the UN Army Group Commander-in-Chief, the provisional armored brigade will advance into Kuwait either (1) along the Dahran-Kuwait City road across the border, or (2) through a known gap in the Iraqi defenses to Al Wafrah and thence to the Dhahran-Kuwait City road. Once past the Iraqi fortified area, the brigade will move as rapidly as possible to Kuwait City, to occupy the city, restore order, and to raise the Kuwaiti flag.

As soon as the provisional brigade reaches Kuwait City, and not before, the Marine Expeditionary Brigade (MEB) will begin landing along the beaches of Kuwait City, and await further orders from the Task Force commander; in the absence of other orders, and if necessary, the MEB will support the Provisional Armored Brigade as necessary, in the event it should encounter opposition.

Other Courses of Action

It is, of course, possible that well-dug-in, disciplined soldiers will not only survive the pounding received in Phase Two of Operation "Colorado Springs," but will retain substantial fighting capability. If so, one or more of the other courses of action will be required. But, should this be the case, the damage done by Operation "Colorado Springs" combined with continuing devastating air support of the attacking ground troops, will make the tasks of the soldiers much easier than they would otherwise have been.

In the event that there is no Iraqi collapse, the UN Combined Forces Commander-in-Chief will issue the necessary orders to initiate one of the other courses of action. We shall examine these in subsequent chapters.

CHAPTER 6

Option Two: Operation "Bulldozer"

Upon the outbreak of hostilities we have assumed that the UN forces in Saudi Arabia will be formed in two armies. On the right will be the UN Eastern Army, consisting of the United States Army's XVIII Airborne and VII Corps, and a provisional corps from the US Marine Corps. This army would be located south of the Kuwaiti-Saudi Arabian border, to include the coastal region where most of the US Marine forces are positioned. The commander of the Eastern Army will be an American general.

The UN Western Army will be made up of the various contingents from the Arab Allies, and would extend westward parallel to the Iraqi-Saudi Arabian border. The commander of the UN Western Army will be a Saudi Arabian general.

Overall ground commander will be the UN Army Group Commander-in-Chief. He will have available the Anglo-French corps in Army Group reserve.

For obvious reasons a flanking maneuver which avoids a direct assault on the main Iraqi defenses in southern Kuwait would be preferable to a direct assault. A frontal attack on these defenses would require considerable application of firepower and manpower to push the Iraqis out of their defensive positions; hence the name "Bulldozer."

Such a course of action would be less risky operationally, though expensive in terms of casualties, since UN infantry and armor would most likely have to become involved in close and bloody fighting to drive the Iraqis out. It is because of the potential cost in UN casualties that "Bulldozer" may be a less desirable course of action than other possibilities.

There could, however, be circumstances in which a "Bulldozer" operation could be realistic and desirable. For instance:

1) When Iraqi forces first invaded Kuwait, they began establishing defensive positions along the Kuwaiti-Saudi Arabian border. As the months passed the Iraqis not only strengthened these positions in southern Kuwait, they have extended the field fortifications westward more than 100 kilometers along their border with Saudi Arabia.

Because of the extension of the Iraqi defensive line to the west, it may no longer be feasible, or practical, for UN forces to turn the Iraqi flank. Then "Bulldozer" may the only other solution if a ground offensive is required.

2) Should the preliminary air bombardment of the Iraqi positions be so devastating that collapse appears imminent, then a rapidly-conducted "Bulldozer" may hasten the collapse.

3) Another circumstance for a "Bulldozer" operation would be the problem of coordination between the varied and diverse UN commands located in Saudi Arabia, already discussed in Chapter 3. Furthermore, there may be some question as to whether all of the Arab allies will be willing to embark on offensive operations into Iraqi territory. A direct frontal assault may be required, carried out by forces which will be prepared to attack when and where directed.

Such an operation would almost certainly be carried out primarily by American forces, if only because of the overwhelming US military presence and firepower preponderance in the Gulf. The other allied forces in the Western Army would, however, be expected to conduct secondary attacks to pin down Iraqi forces and otherwise maintain pressure along the Iraqi border.

As with any possible ground scenario in the Persian Gulf, the massive UN air offensive discussed in the previous chapter would precede any ground operations for a period of several days, perhaps a week or longer. As pointed out earlier, this air offensive would focus on Iraqi military targets in Kuwait as well as Iraq. The targets in the second phase of the air campaign would include known Iraqi defensive positions along the Kuwaiti-Saudi Arabian border. Nevertheless, there is no assurance that the air attacks will destroy the effectiveness of these field fortifications.

To assault these formidable Iraqi defenses, the UN firepower superiority must be used to its fullest potential. Therefore, it is likely that, at the request of the army group commander-in-chief, the UN air commander would order "carpet bombing" of sectors of the Iraqi forward defense line through which one or more breakthroughs are planned.

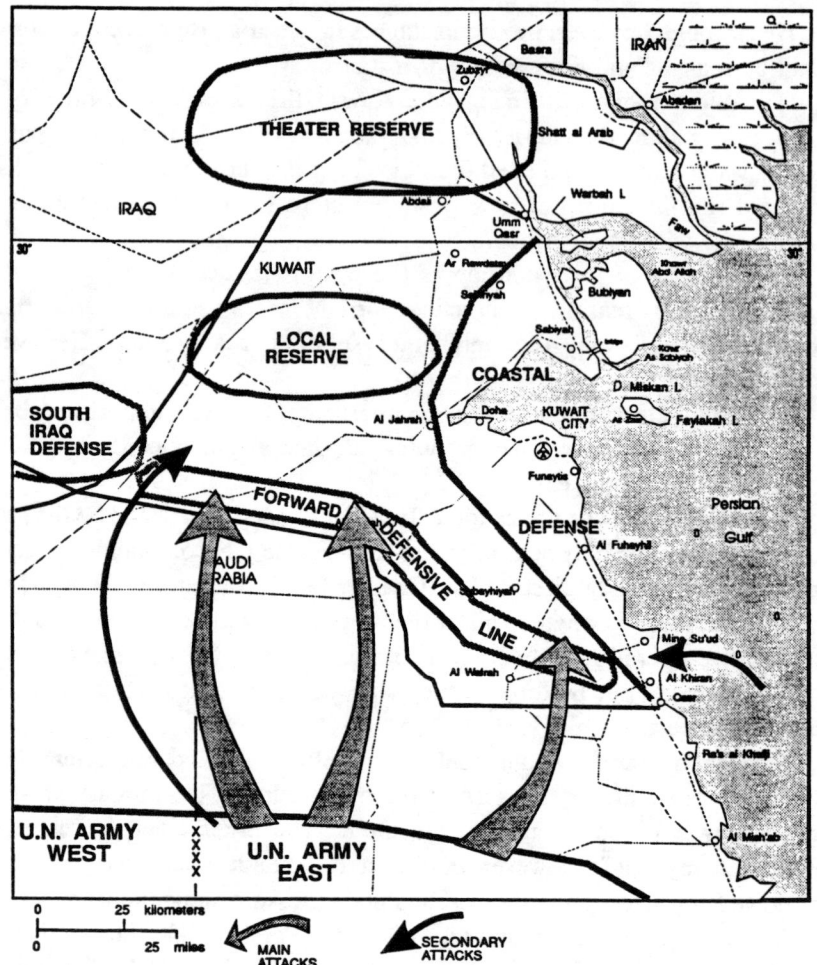

FIGURE 13. OPERATION "BULLDOZER"

Carpet bombings were conducted in World War II on several occasions to open gaps in heavily fortified positions which the ground forces could then exploit. Carried out largely by strategic bombers, carpet bombing attacks focussed on a designated segment of the enemy's field fortification where a breakthrough was desired. The selected area was saturated with heavy bombs which not only had the effect of destroying much of the defensive field works, but also of so demoralizing the surviving defenders that the attacker's ground forces could move through the flattened positions relatively easily. Similar use of this technique was made more recently with the B-52 Arclight raids in Vietnam, with considerable success.

In general the theory of using strategic bombers in such a tactical role proved successful, though on some occasions the result was as devastating to friendly forces as to the enemy, when the bombers missed their mark. Due to improved communications between air and ground forces, however, this is less likely to occur in the coming operations in Kuwait and Iraq.

The concept of carpet bombing would be viable in a UN attack on the Iraqi defenses of southern Kuwait. B-52s from Diego Garcia, augmented by F-111s or A-6Bs, could saturate a designated section of the Iraqi forward defensive line, pulverizing trenches, bunkers and communications lines. Minefields and wire entanglements would also be eliminated. A swathe of destruction can be carved out of the Iraqi defenses in which few enemy soldiers will survive unharmed. Those who do will be too stunned to put up effective resistance. UN ground forces could then begin their assault against the Iraqi forward defensive line. Undoubtedly, bombings of such a magnitude would create large craters which armored and unarmored vehicles and dismounted personnel would have difficulty negotiating. Therefore, engineers will have to be prepared to clear paths for the UN tanks and infantry fighting vehicles. Several such coordinated air/ground assaults can be expected to result in multiple penetrations, leaving many of the surviving Iraqis encircled, and forcing the others either to withdraw or surrender.

To facilitate the advance, as well as to hasten the Iraqi collapse, one or more air assaults would be made by airborne or air assault troops into the area just beyond the carpet bombing. These troops would hold the gaps open, and block any counterattack efforts by Iraqi reserves.

The Iraqis were often defending during their eight year war with Iran. This gave them opportunities to perfect their defensive tactics, and thus was a major contribution to their eventual victory. Iraqi defensive doctrine is built upon the use of a triangular defensive complex of mutually supporting positions or strongpoints.

Beginning at the division level, each triangular set of positions is composed of smaller defensive triangles at each lower level from brigade through battalion and company. The base of the triangle faces the likely direction of attack, with the apex point to the rear.

At the company level, there are machine-gun positions in the triangle's two forward points, while the company's heavy weapons, such as mortars and more potent antiarmor weapons, are located at the apex. Iraqi infantrymen are positioned in trenches or foxholes connecting the triangle's three strong points.

Option Two: Operation "Bulldozer"

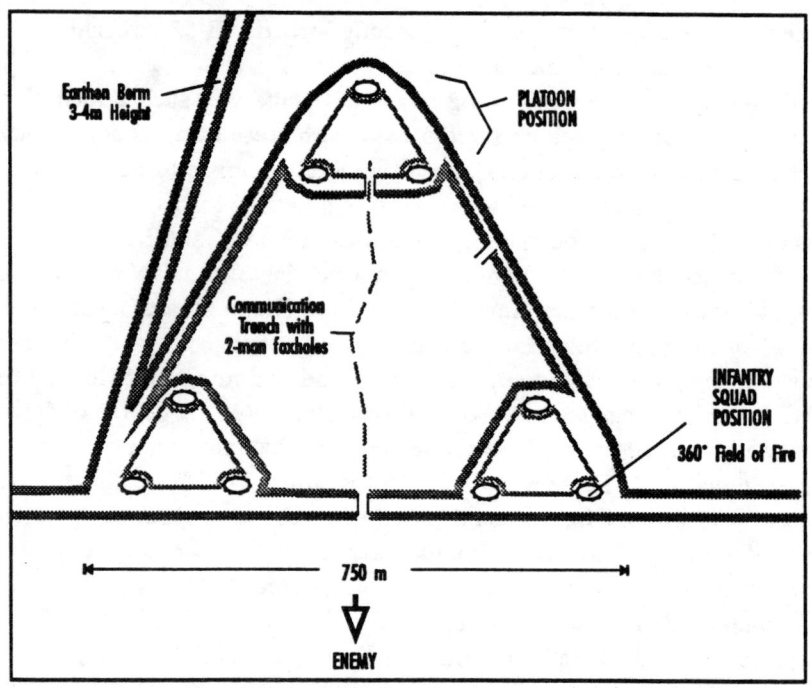

Figure 14. Iraqi Triangular Defensive Position

Iraqi tanks are often dug in on both flanks of the triangle, while to the front there are antitank ditches (some filled with flammable liquids), as well as mines and concertina barbed and razor wire. Armored and mechanized reserve forces are located a short distance to the rear to allow for rapid reinforcement and counterattack.

The Iraqis lay out their minefields so as to require the attacker to concentrate his armor into an open "kill zone," where Iraqi antiarmor weapons can be used to greatest effect. The tanks located on the flanks of the triangle can also get flank shots on the attacking armored vehicles as they move into or through the kill zone.

Each Iraqi infantry squad has at least one RPG-7, hand-held antitank rocket, to be used in close action (500 meters or less) against enemy armor. At the rearward tip of the defensive triangle, mortars, recoilless rifles, or ATGMs (Anti-Tank Guided Missiles) engage the attacker at longer ranges.

Overrunning a resolute enemy in such a defensive position will require both skill and considerable firepower. The best time to attack will be at night. Not only is there likely to be less interference from sandstorms and surface dust, but the Iraqis' night combat capabilities are much less effective than those of the UN Allies. The US Army, for instance, has better night vision devices and more of them than the Iraqis. During the Iran-Iraq War, the Iranians noticed a substantial decline in Iraqi capabilities and morale at night and because of this often conducted their operations after dark.

As mentioned earlier, any ground operation would be preceded by an air offensive, which would have done at least some damage to the Iraqi forward defensive line. Regardless of the strength of such air attack it is assumed that some Iraqi defenders will be left capable of effective defense. These defenders will have to be incapacitated in one way or another.

If we assume that an Iraqi triangle survives the air and subsequent artillery preparation relatively intact, it will have to be taken by combined armor and infantry assault. This will have to be mounted quickly and decisively, preferably under the cover of darkness.

Under direct covering fire from assault tanks, mine-clearing tanks (pushing large rollers to detonate mines) will clear corridors through the minefields, while other armored engineer vehicles equipped with bulldozer blades on the front, will fill in the antitank ditches. (It is unlikely that the ditches filled with flammable liquids would be more than a nuisance. In the first place, it is doubtful if the storage containers will survive the preliminary air and artillery bombardments, and even if they do, the liquid will soon seep into the sand, extinguishing the flames.)

With avenues opened through the minefields and ditches, fast moving M1A1 Abrams tanks and Bradley fighting vehicles will push through the heart of the triangle and assault the heavy weapons located at the tip. The Iraqi reserves, if they live through the air and artillery preparations, will then be engaged by the tanks. The Bradleys and their accompanying infantry will pin down and eliminate the Iraqi infantrymen and RPG-7 gunners stranded in the trenches.

Meanwhile, forward observers (FOs) and air liaison officers (ALOs) will be directing mortar, artillery, and air strikes against particularly stubborn defenders; alternatively, or these can be bypassed to be dealt with by following forces. UN armored forces, having penetrated into the Iraqi defenses, will then move on to the next triangle. As more of these positions are taken, the Iraqi defenses will eventually collapse for lack of mutual support.

Concurrent with the UN frontal assault on the Iraqi defensive positions, additional UN forces will be transported by helicopter to set up blocking positions behind the Iraqi lines, and just in front of the friendly spearheads. There they will obstruct any Iraqi attempts to reinforce or counterattack a threatened sector. These UN forces will have to be amply armed with antiarmor weapons and must maintain close contact with their artillery and air support in order to defend their positions effectively.

A rapid linkup between the two forces will be vital to the success of the operation, assuring a breakthrough. Waiting UN armored forces can then exploit the gap, fanning out to attack remaining defenders from the rear, and causing a collapse of all or a part of the Iraqi defensive line.

With the Iraqi defenses along the Kuwaiti-Saudi Arabian border behind them, UN armored forces must then be prepared to engage Iraqi reserve forces in central Kuwait. These forces may contain elements of the Republican Guards which will be moving southward from their positions in southern Iraq to counterattack the UN forces. These Iraqi reserves may be heavily attrited due to the effects of "Colorado Springs"; large numbers may have survived, however, and remain a viable threat.

While the UN ground forces are pushing through the Iraqi defenses of southern Kuwait, naval gunfire and missiles, as well as carrier-based naval and Marine air, will pound any coastal defenses that have survived the air attacks in Phase Two of "Colorado Springs."

It is at this point that Operation "Salah-al-Din" may be initiated in which a combined Kuwaiti-Saudi Arabian armored force would dash into Kuwait City.

On to Baghdad?

The objectives of the Un Security Council resolutions will be accomplished when (1) all of Kuwait is liberated and secured, (2) Saddam Hussein (or his successor) no longer has a capability for new aggression against other neighbors, and (3) action has been initiated to assure reasonable restitution to Kuwait for the terrible damages that have been inflicted on that small nation. Implicit in all of these resolutions, and particularly in Resolution 678 (which authorized the use of military force after 15 January), is a fourth result: existence in Baghdad of a law-abiding, peaceable regime, capable of concluding with the United Nations whatever agreements are necessary to terminate the crisis precipitated by the Iraqi invasion of Kuwait on 2 August, 1990.

The first two of these results will obviously be achieved by the kind of military action described in this and the previous chapter. The third result (some kind of program of restitution or reparations) is clearly dependent upon the fourth: a responsible government, exercising authority over Iraq.

There are two principal ways in which the military operations will be terminated. One is a formal surrender or ceasefire negotiated either with the current regime, or with a successor regime which has overthrown or otherwise replaced Saddam Hussein's government. The other is simply the collapse of the Iraqi armed forces, resulting in a *de facto* ceasefire.

In the second of these cases, the country will probably be in chaos, without an effective central government. This virtually demands that the UN forces occupy Baghdad, and possibly all or most of the other major population centers in the country. In the event of a formal surrender or ceasefire, it may or may not be necessary or desirable for UN forces to occupy Baghdad, either to support, or to dominate, the government.

Obviously, therefore, the UN Combined Forces Commander-in-Chief must have plans for an advance by some part of his forces to occupy Baghdad. Depending on the circumstances, such a movement might or might not encounter opposition.

CHAPTER 7

Option Three: Operation "Leavenworth"

The United States Army Command and General Staff College, located at Fort Leavenworth, Kansas, is where the Army trains its most promising officers to become staff officers and commanders of large combat formations. It is also where the tactical and operational doctrine of the US Army is formulated. Some people believe that the professional excellence of this institution in the 1920s and 1930s—when it was known as a "School" and not a "College"—was largely responsible for the amazingly successful performance of the US Army in World War II, while undergoing a 100-fold expansion. Not all "Leavenworth men" get to the top in the Army. But there have been few top Army commanders of the past 50 years who have not been Leavenworth men.

When the average American civilian hears the word Leavenworth, he is likely to think of a notorious Federal penitentiary. In the armed forces, however, the word Leavenworth means both a fountainhead of military learning and military academic excellence. A "Leavenworth solution" is usually related to the tactical problems which the students have to solve during their course of study. It is considered that "the approved solution" of one of these problems is indicative of how Napoleon would have solved the problem, had he been a student.

Most people have heard the sick old joke about "the right way, the wrong way, and the Army way." Army professionals believe, however, that the right way and the Army way are both the Leavenworth way.

It is very doubtful if the concept for Operation "Bulldozer" would ever be an approved solution at Fort Leavenworth's Command and General Staff

College (C&GSC). Despite some well-publicized theories about "the American Way of War," Leavenworth has never endorsed a concept which some critics of the Army have referred to as the Army's preferred method of "attrition warfare." And the "Bulldozer" concept is unadulterated attrition warfare. The Leavenworth graduate will always seek to achieve the greatest possible firepower superiority, but he is also indoctrinated with the concept of applying imagination and maneuver—and a certain amount of caution—to the solution of a battlefield exercise.

A student can receive one of two possible grades for his solution to a tactical problem at Fort Leavenworth: "S" or "U." An "S" is a Satisfactory Solution—although not necessarily *the* "approved solution." A "U" means an Unsatisfactory Solution: a failure. The concept presented in the previous chapter for Operation "Bulldozer" would very possibly earn a "U" if it were the solution to a problem at Fort Leavenworth.

Concept of "Fire and Movement"

The staff of the UN Army Group has available for a ground offensive a total of 18 divisions in seven corps in two armies, as follows:

The ground forces available to the commander of the Eastern Army are the following:

3 corps (VII, XVIII Airborne, Marine), consists of:
- 1 US airborne division
- 1 US air assault division
- 2 US mechanized infantry divisions
- 4 US armored divisions
- 2 US Marine divisions (plus 2 Marine Expeditionary Brigades afloat)

The ground forces available to the commander of the Western Army are the following:

3 corps (Saudi, Egyptian, Syrian), consists of:
- 2 armored divisions (Egyptian, Syrian)
- 4 mechanized infantry divisions (2 Saudi, 1 Egyptian, 1 Arab provisional)

In army group reserve, directly under the command of the UN Army Group Commander-in-Chief, is the Anglo-French provisional corps, with the British 1st Armored Division, and the French 6th Light Armored Division.

Thus the grand total of UN land force troops is seven corps headquarters and 18 divisions.

OPTION THREE: OPERATION "LEAVENWORTH"

The most powerful, most mobile, most cohesive major force is the Eastern Army, which will make the main effort in the planned offensive. The Western Army, therefore, will make a holding attack to pin down the Iraqi forces of the South Iraq Defensive Force, so they cannot interfere with the Eastern Army's main attack against the Iraqi South Kuwait Defensive Force.

The VII Corps will be deployed just south of the Saudi Arabia-Kuwait border, with the mission of making a secondary, or holding attack against the front line defenses of the South Kuwait Defensive Force. The corps will have the 1st Mechanized Infantry Division, the 24th Mechanized Infantry Division, and the 1st Cavalry Division in line, with the 3d Armored Division in corps reserve.

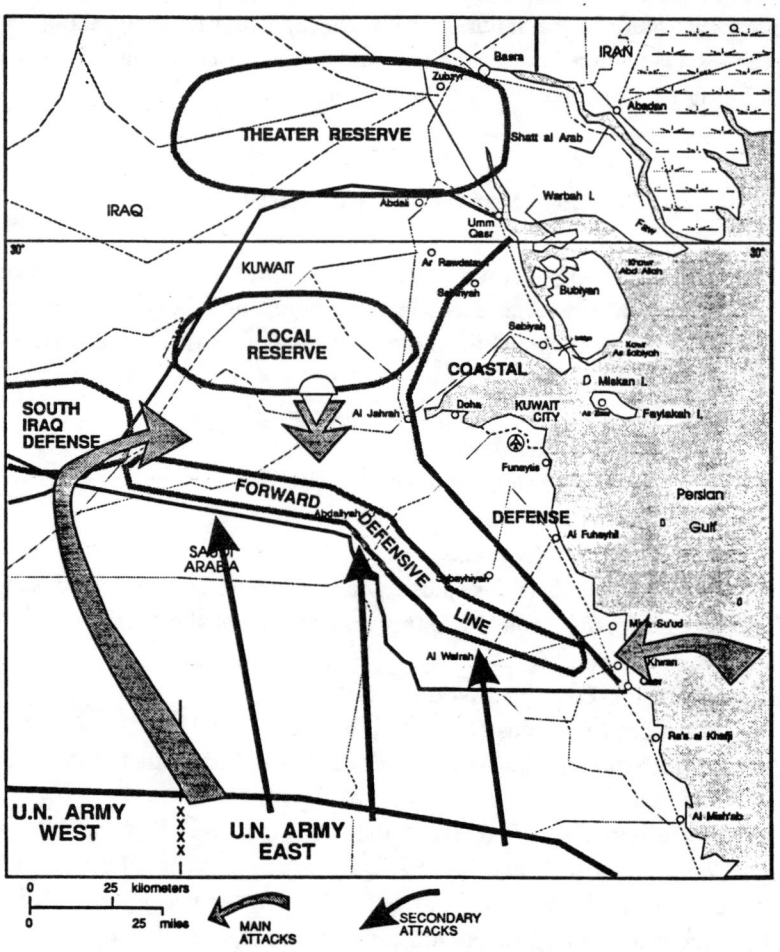

FIGURE 15. OPERATION "LEAVENWORTH"

OPTION THREE: OPERATION "LEAVENWORTH"

The XVIII Airborne Corps will make the main effort, on the left, spearheaded by the 1st Armored Division, followed by the 2d Armored Division. While the 1st Armored Division, with massive air support, is making a breakthrough, just west of the Iraq-Kuwait border, the 82d Airborne Division will make a parachute and air assault landing just behind the right flank of the South Kuwait Defensive Force. This division will be followed by the 101st Air Assault Division, making an airlanding inside the airhead established by the 82d Division. While the 101st Airborne Division establishes a blocking position to the north, in the event of a counterattack by the Iraqi Central Kuwait Reserve Force, the 82d will attack south. It is anticipated that within 36 hours the 1st Armored Division will pass through, to roll up the right flank of the South Kuwait Defensive Force. The 2d Armored Division, following the 1st, will attack southward, or southeastward, depending upon the situation at the time it passes through the 101st and 82d Divisions' airhead.

Meanwhile, on the right the 4th Marine Expeditionary Brigade (MEB) will make an amphibious assault landing in the general vicinity of Mina Sa'ud, to hit the right flank of the South Kuwait Defensive Force. Elements of the 4th MEB will drive south along the Kuwait-Dahran Road to link up with the right flank of the VII Corps. Once this linkup is achieved, the 1st Marine Division followed by the 2d Marine Division (in Eastern Army Reserve) will advance northward up the Dhahran-Kuwait Road, toward Kuwait City, being supported by both the 2d and 3d Marine Air Wings and naval gunfire and missile support.

In the likely event that the double envelopment by the XVIII Airborne Corps and the Marine Provisional Corps creates disruption and demoralization in the front lines of the Iraqi South Kuwait Defensive Force, the VII Corps will shift from a holding attack to an assault role. As soon as one of the three front-line divisions achieves a breakthrough, this will be exploited by the VII Corps reserve, the 3d Armored Division. This division will drive toward the Kuwait International Airport.

Depending on the general state of continuing Iraqi resistance, the UN Army Group Commander in Chief will consider initiating Operation "Salah-al-Din," for the liberation and occupation of Kuwait City, as described in Chapter 5. He will also need to consider the possibility of occupying Baghdad, also as discussed in Chapter 5.

CHAPTER 8

Option Four: Operation "RazzleDazzle"

A Napoleonic approach to the situation facing the UN forces in northern Saudi Arabia and the Persian Gulf would almost certainly be bolder and more dramatic than Operation "Leavenworth." (That is, if the logistical situation permits it.) This approach would be designed to exploit to their utmost the special capabilities of the forces available: As with "Leavenworth," the ground forces are the following:

For the Eastern Army:
   3 corps (VII, XVIII Airborne, Marine)
     1 US airborne division
     1 US air assault division
     3 US mechanized infantry divisions
     3 US armored divisions
     2 US Marine divisions (plus 2 MEBs, afloat)

For the Western Army:
   3 corps (Saudi, Egyptian, Syrian)
     2 armored divisions (Egyptian, Syrian)
     4 mechanized infantry divisions (2 Saudi, 1 Egyptian, 1 Arab provisional)

In army group reserve there are two more divisions, both armored, one British and one French, both lighter than the US armored divisions, in a provisional Anglo-French corps. Thus the total in division types is as follows:
   1 airborne
   1 air assault
   7 armored
   7 mechanized infantry

Option Four: Operation "RazzleDazzle"

2 Marine (amphibious-capable) divisions
2 Marine Expeditionary Brigades (afloat)

The "RazzleDazzle" Concept

Bearing in mind the political problems discussed in Chapter 2 and the command problems discussed in Chapter 3, the Combined Forces Commander-in-Chief and the UN Army Group Commander-in-Chief must both have doubts as to the extent to which they can rely upon the Western Army for offensive maneuvers into Iraq. Without writing off the possible utility of those six divisions for a secondary or holding attack, the army group commander-in-chief has available for offensive maneuvers the following division types in the Eastern Army and in the UN Army Group Reserve:

1 airborne
1 air assault
5 armored
3 mechanized infantry
2 Marine divisions
2 MEBs (afloat)

The very nature of these forces, combined with the nature of the terrain, and the static defenses, suggests the desirability of encirclement of the three Iraqi corps in Kuwait by means of a triple envelopment: a standard double envelopment (partly amphibious) combined with a vertical envelopment, as follows:

- A wide turning movement from the desert in the west by armored forces driving eastward into northwestern Kuwait between the Local Reserve Force in central Kuwait and the Strategic Reserve Force south of Basra;
- Another wide turning movement by amphibious forces from the east driving westward into northeastern Kuwait north of Kuwait City, between the Local Reserve Force in central Kuwait, and the Strategic Reserve Force south of Basra;
- An airborne and airlanded assault into north-central Kuwait, northwest of Kuwait City, between the Local Reserve Force in central Kuwait, and the Strategic Reserve Force south of Basra;
- A vigorous holding attack from the Eastern Army sector in northern Saudi Arabia to hold the Southern Defensive Force in place;
- Another vigorous holding attack by the Western Army into southern Iraq, west of Kuwait, to hold the South Iraq Defensive Force in place.

Let's look further at each of these plan components.

Option Four: Operation "RazzleDazzle"

Turning Movement from the West

The XVIII Airborne Corps will organize two two-division task forces for this operation: a Desert Task Force and an Airborne Task Force. The Desert Task Force will consist of the 1st Armored Division and the 2d Armored Division. This task force will have the mission of either getting through, or behind, the South Iraq Defensive Force to drive into northwestern Kuwait from the west.

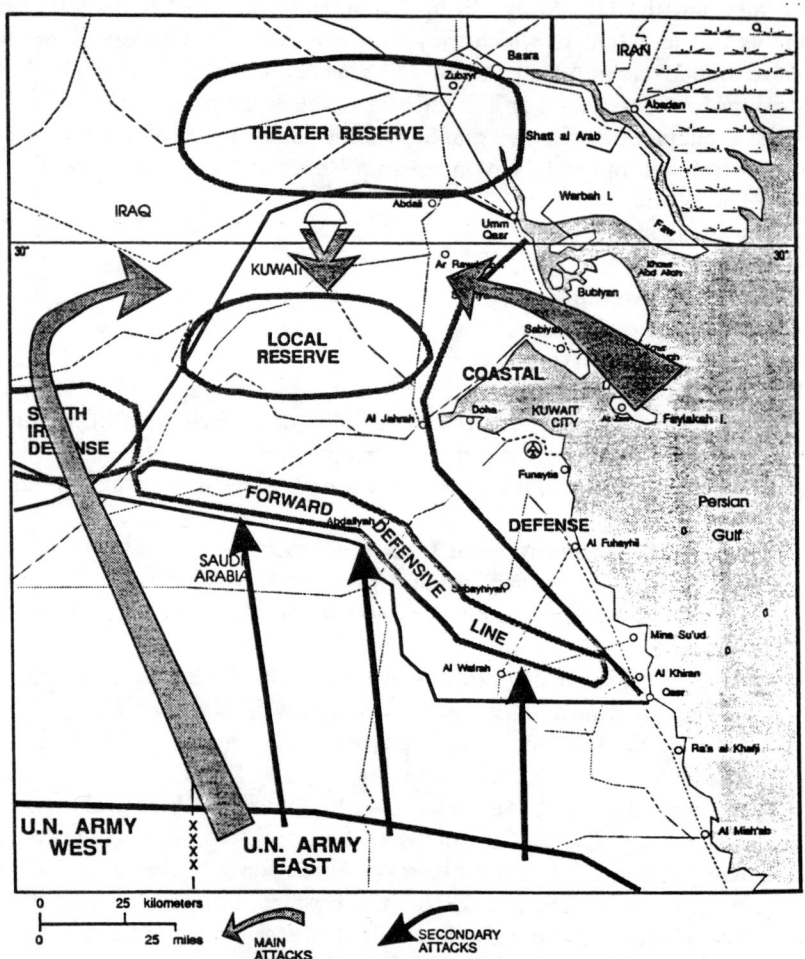

FIGURE 16. OPERATION "RAZZLEDAZZLE"

One possible route will be to move behind, and to the west of the Western Army, and thence north, generally through Nisab, about 50-75 kilometers into Iraq, to the vicinity of Juhaym, then turn east, behind the South Iraq Defensive Force, to cross the Kuwait border just south of 30 degrees N. Latitude. This would be a long march, approximately 300 kilometers, and could be considered only if the previous air campaign had assuredly impaired the capability of the South Iraq Defensive Force to interfere. Logistic sustainment of such a move would also be a serious problem.

The alternative route would be to start by blasting a gap through the eastern positions of the South Iraq Defensive Force, just west of its junction with the South Kuwait Defense Force, near the western border of Kuwait. (This would be similar to the preliminary move in the main effort for Operation "Leavenworth.") Once the gap is opened, the 1st Armored Division would drive northeastward, again to cross the Kuwait border just south of the 30th Parallel of Latitude.

Once inside of northwestern Kuwait, the 1st Armored Division would establish a blocking position, facing generally north, toward Basra, to prevent interference from that direction by Republican Guard units of the Iraqi Strategic Reserve Force. The 2d Armored Division, following behind the 1st Armored, would deploy, facing southward, behind the blocking position of the 1st Armored Division, and would then attack southward into the rear of the positions of the Local Reserve Force.

Turning Movement from the East

The turning movement from the east will be directed generally across Bubiyan Island into the Sabiriyah and Ar Rawdatayn oil fields on the mainland. This operation will be undertaken by the 1st Marine Division, with the 4th MEB and 5th MEB attached. The offensive will be spearheaded by the (to which the two air wings of the 1st MEF will be attached for the assault phase). Bubiyan Island is currently held by an Iraqi Marine brigade. A preparatory air and naval gunfire and missile fire bombardment against the coastal positions of this brigade will be initiated 24 hours before H-Hour (at about the same time as the 1st Armored Division is driving into southern Iraq further west). Simultaneously with the initiation of this bombardment elements of the 5th MEB will assault and seize Miskan Island and the western coast of Faylaka Island, just south of Bubiyan Island. As soon as the selected areas are secure, medium and heavy artillery will be put ashore to provide long-range support for the main assault the following day.

The landing on the southern portion of Bubiyan Island, the following day, will secure only the southwestern portion of the island, to provide a base for a following assault crossing across the Khawr As Sabiyah Channel. If the existing bridge has not been demolished, it will be seized and a bridgehead established on the Sabiyah Peninsula on the mainland. In any event, engineers will promptly construct other bridges across the channel, to enlarge the bridgehead, and permit a further offensive westward and northwestward into the oil fields. Leading this second amphibious assault will be the 5th MEB which will secure the Sabiriyah oil field area, and form a blocking position facing north against possible interference from Republican Guard units of the Strategic Reserve Force. If active patrolling indicates that there is no resistance in the Ar Rawdatayn oil fields to the northwest, they will also be seized, and the blocking position pushed further north.

Meanwhile the 1st Marine Division will follow the two MEBs ashore on Bubiyan Island and onto the Sabiyah Peninsula beachhead. The division will advance west to the Basra-Jahrah highway, and prepare to attack southward.

The Vertical Envelopment

As soon as both the 1st Armored Division and the 1st Marine division have moved into mainland Kuwait, the 82d Airborne Division will commence an airborne and air assault landing about 50 kilometers northwest of Jahrah, west of the Basra-Jahrah highway. An airhead will be established, and the Airborne Task Force command post will be set up. The 101st Air Assault Division will be landed in the airhead as quickly as possible. The 82d Division will then advance southward, establishing contact with the 2d Armored Division on its right, and the 1st Marine Division on its left. The Airborne Task Force commander will then assume command of all three divisions for an attack south into the positions of the Iraqi Local Reserve Force.

At the same time, to the north, the Desert Task Force headquarters (which already has the 1st Armored Division under its command) will also assume command of the 10st Airborne Division and the 4th and 5th MEBs. This task force will have the mission of preventing any interference by the Iraqi Strategic Reserve Force. Active patrolling will be undertaken toward Umm Qasr and Basra, to ascertain the status of the Strategic Reserve Force, following the air bombardment. If resistance is light or negligible, those two cities will be seized. If resistance is heavy, the Desert Task Force will dig in, reorganize for defense, hold its positions, and await further orders.

Meanwhile the southward advance of the Airborne Task Force will be pressed vigorously. If resistance is light, the advance will be continued until contact is established with VII Corps units. In this case, the 1st Marine Division will advance to the western outskirts of Kuwait City and halt.

Operation "Salah-al-Din"

The arrival of the 1st Marine Division at the outskirts of Kuwait City will be a signal to the UN Combined Forces Headquarters to initiate Operation "Salah-al-Din." (Named after the great Arab sultan of Egypt who reconquered Jerusalem from the Crusaders in the 12th Century.) This will be done slightly differently than in the concluding phases of Operations "Colorado Springs," "Bulldozer," and "Leavenworth."

Presumably by this time the VII Corps will have begun an advance into the crumbling Southern Defensive Force positions. If, indeed, the defense is crumbling at this time, the VII Corps will be directed to send its reserve division—the 3d Armored—north past Wafrah toward Kuwait City Airport. Simultaneously the army group reserve—the Anglo-French Corps—will be directed to pass though the center or left of the VII Corps, west of Abdalliyah, and also in the general direction of the airport. (The boundary line between the Anglo-French Corps, on the left, and the 3d Armored Division, on the right, would be a line from Abdallyah to the airport.)

The Salah-al-Din Task Force would be committed to follow either the 3d Armored Division, or the Anglo-French Corps, depending upon the status of resistance. If possible it would follow the advance of the 3d Division, since this is the best road route to Kuwait City. As soon as the Anglo-French Corps and the 3d Division reach the line Jahrah-Airport-Funaytis, they will stop, and allow the Salah-al-Din Task Force to pass through.

As soon as it receives word that the Salah-al-Din Task Force is north of the Airport, the 1st Marine Division will advance slowly eastward into Kuwait City. Patrols will seek to establish contact with elements of the Task Force. The division's mission will be to facilitate the occupation of Kuwait City by the Salah-al-Din Task Force, and to provide support to that Task Force as required. Should serious resistance be encountered, the 3d Armored Division and the Anglo-French Task Force may also be called upon to provide support.

Again, the UN Combined Forces Commander-in-Chief will probably need to consider the possibility of occupying Baghdad.

Assessment

Napoleon would probably approve the "RazzleDazzle" concept outlined above. At Leavenworth it would certainly rate an "S-X" (Excellent Satisfactory Solution).

CHAPTER 9

Jordanian Diversion

As pre-hostilities pressure mounts against Saddam Hussein, both militarily and economically, he will continue to seek ways to break up the anti-Iraq alliance by capitalizing on the general Arab hostility toward Israel. An obvious way to do this would be to maneuver Israel into conducting military operations against Iraq. The rationale for this is the assumption that it would be intolerable to the other Arab participants in the UN alliance to be fighting on the same side as Israel against another Arab nation, even a pariah Arab nation. Saddam would hope that these Arab nations, now aligned against Iraq, would then shift to an anti-Israel, pro-Iraq stance.

One way of provoking an Israeli reaction would be to fire long-range Scud missiles into Israeli territory, presumably targeted on Tel Aviv, or Jerusalem, or both. Saddam might do this either in anticipation of a UN offensive to liberate Kuwait, or (as he has threatened) as an initial response to such an offensive. Should he do this as a pre-hostilities pre-emption, he cannot be certain that the Arab members of the UN coalition would react as he would like. They might be so disgusted by the unprovoked attack on civilian populations as to ignore (while secretly applauding) a massive Israeli counterstroke. If the missile attacks on Israel were to be a response to a UN offensive, presumably Israel could be persuaded by the United States not to respond, and to leave retribution to the UN allies.

A more likely way for Saddam to provoke Israeli military action, or threat of action, would be for him to move into Jordan. Presumably this would prompt an immediate Israeli invasion of Jordan to protect her eastern border, and to occupy Jordanian airfields through which Iraq might stage air strikes against Israel. This would be risky for Iraq, since Israel would probably also accompany such movement with air strikes against critical military targets in Iraq. Saddam presumably would believe this risk would be tolerable if Israel were thus committed as an active belligerent, since he would expect that most, if not all, of the UN Arab allies would shift sides to come to his aid.

Five major questions must be addressed in considering how events would unfold if Saddam Hussein were to initiate such a scenario:
1) How would Jordan react to the Iraqi move?
2) Would the Israelis react in the fashion expected?
3) If Israel did react as expected, how seriously could or would Jordan interfere with such an Israeli action?
4) What other options do the Israelis have?
5) What is the likely reaction of the United States Government?

Jordanian Reaction

An unprovoked and uninvited Iraqi military movement into Jordan would render even more difficult the near-intolerable political situation of Jordan and its king. The excellent, but small, army of this impoverished, weak, and internally-divided kingdom cannot repel a serious invasion from either Iraq or Israel. King Hussein has paid lip service agreement to the UN sanctions against Iraq, while at the same time retaining a cordial, but arms-length, relationship with Iraq's unrelated President Saddam Hussein.

Most (but not all) of the Palestinian majority of King Hussein's country supports Saddam Hussein and Iraq because they see him as an enemy of their enemy (Israel), and the potential leader of a united "Arab Nation," who will destroy Israel and return them to their homeland in Palestine, now occupied by he state of Israel. Most (but not all) of the minority Bedouin and native Transjordan population seem to be more in agreement with the Arab opponents of Saddam Hussein than they are with his transparently less than sincere protestations of support for Arab and Islamic ideals and objectives.

The army is probably more in sympathy with this minority than with the Palestinians. Thus King Hussein is "damned if he does and damned if he doesn't." It would be difficult for him to invite Iraqi intervention in Jordan unless his regime were to be threatened by civil war, popular uprising, or military mutiny.

It is likely, therefore, that Saddam Hussein will create a situation prompting an appeal for help from King Hussein. The best way to do this would probably be by stirring up such serious Palestinian unrest as to threaten the overthrow of the monarchy. On the other hand, this would take some time, and would be likely to alert his enemies to his intentions. He might, therefore, risk the opposition of the Jordan army by a surprise invasion.

Israeli Reaction

It appears probable that the Israelis would react to an Iraqi invasion of Jordan in the manner expected by Saddam. Such a reaction would have a defensive as well as offensive component.

Defensively, Israel can be expected to occupy most of Jordan west of the Mafraq-Zarqa-Ma'an-Aqaba highway, and all airfields that the Iraqi air force might wish to use to stage attacks against Israel, particularly Mafraq and Azraq.

Offensively, the Israelis can be expected to strike all or most Iraqi airfields west of the Euphrates River, all Scud launching sites in range of Israel, and possibly a number of selected military depots and Iraqi industrial plants, particularly those involved in chemical, biological or nuclear weapons production.

The Jordanian Army, though well trained, cannot stop an Israeli ground invasion, although it could seriously disrupt it. However, the Israelis, probably assuming that the Jordanians would be deployed for the defense of Amman, will in all likelihood seek to avoid an initial head on battle with the Jordanians. Except for likely air strikes at command facilities and airfields in the Amman area, they will probably avoid Amman, and within 72 hours be in position on the highway line, with an advance airborne brigade holding Azraq. They will then probably seek a cease-fire with the Jordanians. Should events take this course, the United States might be able to negotiate such a cease fire.

Israeli and Iraqi Interaction

Probably the only basis upon which Israel would agree to delay such a reaction would be a United States guarantee to take whatever military action is necessary to prevent the Iraqis from reaching Mafraq, Zarqa, and Amman.

This would have to be a guarantee that was credible and that the Israelis knew was realistic.

Saddam Hussein would not take such an action lightly, since he could expect that the responses from Israel and the United States would be swift and violent. The Israeli air response alone could be devastating. It would also provide the United States with an unequivocal basis for initiating direct military action against Iraq (particularly in Kuwait) which the US might otherwise be reluctant to initiate unilaterally, even after 15 January.

If, however, Saddam believes that UN forces, or US forces alone, are about to take action under UN resolution 678 (the 15 January ultimatum), he might conclude that it would probably be worthwhile to incur the costs and risks attendant upon provoking an Israeli reaction, in the hope that this would destroy the US-Arab alliance which now sustains the US position in the Middle East.

Should the scenario mentioned previously be played out, with the subsequent collapse of the anti-Iraq Arab League, the result would be disastrous for US interests in the Middle East.

United States' Reaction

It would be impossible for the United States to prevent an Iraqi invasion of Jordan, though Saddam could possibly be discouraged from doing so. Let us assume, however, that the United States has provided Israel with a guarantee that Iraqi forces will not be allowed to occupy central Jordan. In return Israel will agree to refrain from immediate response to an Iraqi move into eastern Jordan. We shall also assume that the US has initiated the necessary plans and preparations to be able to carry out those guarantees in the event Iraq decides to incur the risks of an invasion of Jordan.

Operation "Jordan"

An Iraqi move into Jordan, therefore, will automatically trigger Operation "Colorado Springs," with widespread air strikes against Iraqi military and industrial targets. At the same time Operation "Jordan" will be initiated. This would involve the air interdiction of the highway from H4 and H3 to Baghdad, destroying all possible Iraqi movement on the road. US air forces will also provide cover and support to air assault operations by US ground forces to secure the airfields at Azraq and Mafraq. (US carriers in the Mediterranean and Red Sea would contribute to this.) As soon as these fields are secured UN air units will be stationed there.

Simultaneously with the issuance of orders for Operation "Jordan," the government of Jordan will be notified that action is being taken to occupy the air bases at Azraq and Mafraq, and Amman will be requested to cooperate. If the Jordanian armed forces interfere with the occupation of these two bases, the minimum force necessary will be used to seize, secure, and hold the bases for an indefinite period of time.

The operation most likely will be conducted by either the 101st Air Assault or 82nd Airborne divisions; which are most suited for such an operation. The assault force will be designated "Task Force Jordan." A brigade will be used to seize each air field, with two more brigades being air transported shortly thereafter. If Jordanian opposition is significant, then another airborne or air assault division, reinforced by an armored brigade, will be air transported to the two airheads.

Task Force Jordan will establish an observation line from about 20 kilometers east of Azraq, north to the Syrian border, about 20 kilometers east of Mafraq. It will be prepared to attack any Iraqi forces that get as far west as this observation line.

Israel, therefore, can remain a spectator, as is desired by all other participants, except Iraq.

CHAPTER 10

The Sobering Effect of Logistics

Logistics can be defined as the art or science of providing armed forces with the materials and services they require to fulfill their mission. Logistics therefore includes supplies, equipment, transportation, maintenance, construction, the operation of facilities, and the care, movement, and evacuation of wounded, ill, or injured personnel. As the old soldiers' saying goes, "Amateurs talk strategy; professionals talk logistics." (A modern scholar has suggested that this should be changed to read: "Amateurs talk strategy; professionals know strategy depends on logistics.") Napoleon echoed a similar sentiment in his famous adage that "an army marches on its stomach." Readers interested in the historical background of modern logistics are referred to Appendix F, "Logistics in Historical Perspective."

Modern Logistics

The provision of support to enable units in the field to perform their mission is an exacting discipline. Many commentators and so-called "armchair generals" give it insufficient weight when commenting on operations. Even some commanders (like Rommel in North Africa), who ought to have known better, have willfully ignored or dismissed part of the problem. Sometimes they got away with taking such risks; more often they did not.

Essentially, logistics may be divided into two main parts. First, the production, procurement, or assembly of the required materials and personnel, and movement of those materials and personnel to the theater of operations. Second, distributing those items, materials, and personnel to the units which need them. Although supply is perhaps the most readily recognized component of logistics, other important components include medical care (along with evacuation of casualties), transportation (both for supplies and personnel), maintenance, and the assortment of services a modern army needs (including such varied services as postal delivery, laundry facilities, construction, explosive ordnance disposal, and financial services).

The responsibility for handling logistics matters falls to various support commands. The organization of these commands, and the specific allocation of responsibilities, varies between nations and services. In the US Army, most of these activities fall under the control of the divisional support command (DISCOM) and the corps support command (COSCOM). Each division and corps within the theater of operations controls a support command of appropriate size. In addition to strictly logistical operations, support commands also contain such specialized components as military police units, movement and material control centers, and graves registration teams. In simplest form, the purpose of the support command is to handle all the "rear-area" and "housekeeping" activities and so provide the combat and combat support units with what they need to accomplish their mission.

Supply

The US Army currently recognizes ten classes of supply, while forces of other NATO countries employ a system derived from the older, US Army supply classification system, used in World War II and the Korean War. (See Figure 17.)

There are two types of US heavy divisions: armored and mechanized. The armored division has 17,027 men; its principal combat elements are six tank, four mechanized infantry, and four artillery battalions. The mechanized division has 17,330 men, with four tank, six mechanized infantry, and four artillery battalions. These are considerably larger than Korean War-era armored divisions (16,053 men with four tank, four armored infantry, and four artillery battalions). In sheer numbers of vehicles the disparity is even greater, with the 1953 armored division containing some 3,257 vehicles of all types, against over 5,400 for the 1990 mechanized division. Moreover, the contem-

Old US and Modern NATO Supply Class	Modern US Supply Class	Description
I	I	Subsistence, including rations and gratuitous health and welfare items.
II	II	Clothing, individual equipment, tentage, tool sets and kits, administrative and housekeeping supplies and equipment; includes all equipment, other than principal items, prescribed in authorization/allowance tables and items of supply (not including repair parts).
III	III	Petroleum, oil and lubricants (POL), including fuels of all types, compressed gas, coolants and antifreeze, solid fuels.
IV	IV	Construction materials, including installed equipment and fortification/barrier materials.
V	V	Ammunition of all types, including explosives, chemical, radiological, and special weapons; also fuzes, detonators, rockets, pyrotechnics, and related items.
I	VI	Personal demand items (nonmilitary sale items), available through a post exchange.
II	VII	Major end items: equipment ready for its intended use, such as tanks, artillery pieces, mobile machine shops, vehicles.
II	VIII	All medical supplies, including repair parts peculiar to medical users.
II	IX	Repair parts and components of all types (except medical), including kits, assemblies, etc., required for maintenance of all equipment.
IV	X	Materials for nonmilitary programs such as agricultural and economic development if not included in classes V to IX.

* Chart drawn from following sources: US Army, FM 63-3J, *Combat Service Support Operations—Corps* (Washington, DC: US Dept of the Army, August, 1985), pp 2-25 and 2-28; US Army FM 105-5-1, *Operational Terms and Symbols* (Washington, DC: US Dept of the Army, October, 1985), p. 1-14 and Appendix C.

FIGURE 17: SUPPLY CLASSES*

MODE: DIVISION:	Attack	Pursuit	Reserve
1953 Armored	702 total	351 total	136 total
subsistence	43	45	43
fuel/POL	111	187	46
ammunition	401	70	-
other	147	49	46
1953 Infantry	589	314	111
subsistence	51	53	51
fuel/POL	74	144	21
ammunition	418	73	-
other	46	44	39
1990 Mechanized	2743	2079	557
subsistence	46	48	46
fuel/POL	1401	1746	408
ammunition	1002	175	-
other	294	110	97

* Figures in are drawn from US Army FM 101-10 (op cit), pp 296-302; US Army FM-101-1/2 Staff Officers Field Manual: *Organizational, Technical and Logistical Data Planning Factors* (Vol 2) (Washington, DC: US Dept of the Army, Oct. 1987), pp 2-0 to 2-180.

FIGURE 18: SUPPLY EXPENDITURE PER DAY, IN TONS*

porary division has over six times as many aircraft (125, all of them helicopters) than the 1953 unit, which had one helicopter and 19 other aircraft. Further, one round fired from all of the 1953 division's howitzers would total 1.82 tons, while the same fire from today's division would total 4.93 tons, not including 18 tons from the 9-unit MLRS battery.

As Figure 18 demonstrates, the major increase in supply requirements between the 1950s and the present has been the increase in size and weight of artillery and tank gun ammunition, and in the rate of fuel consumption. This has led to a significant increase in overall supply requirements, and requiring more and heavier ammunition supply vehicles to distribute supplies to front-line units. It has been estimated that the US artillery units committed to "Desert Shield" are capable of firing 500 tons of ammunition in one hour.

Modern armies require water, not only for their personnel, who need it to drink, cook, and wash, but also for their vehicles. Even in temperate climates, a modern US mechanized division would require 4.3 to 7.3 gallons of water per man per day, or 298 to 506 tons per day. Modern US Army figures for operations in a hot, arid climate (such as Saudi Arabia and Kuwait) amounts to 9.4 to 12.4 gallons per man per day, or between 652 and 860 tons per day. These totals do not include any allowance for water for motor vehicles. Whether water requirements could be supplied by local desalinization plants and wells is unclear, although these sources do meet the normal peacetime requirements of the local population.

Production

As far as a war in the Gulf is concerned, the production of supplies ought not to be a problem for the US and its European allies. The situation for some of the Arab allies, especially the smaller Gulf states, may be another matter. Most of them are so small that there is no indigenous arms industry, and their materiel stockpiles (if they have any) are small. The larger armies are not faced with such problems, as they maintain significant materiel stockpiles in their home nations, and most of them produce at least some of their own required equipment and munitions.

There is one major area for concern in this sector, however. Modern sophisticated weapons such as laser-guided antitank missiles (like the Hellfire on US AH-64 Apache attack helicopters) and sophisticated antiaircraft missiles, are not produced in great quantity. Increasing production rates would prove difficult, even in the United States, for several reasons: production lines are limited, as are production lines for major components like complex electronics; the skilled workers who assemble these components and weapons cannot be trained hurriedly; some of these components and weapons require special materials or ingredients whose supplies are limited, or which are required for other vital weapons, or for important civilian-sector items.

This twin problem, of limited initial stocks and low production rates, means that it is possible for US and allied forces to run out of certain items. In the admittedly unlikely event of a long Gulf war (one that lasts, say, for over 60 days), stocks of some types of technologically complex munitions could be drawn down so far that field use would have to be severely restricted. Even with restrictions in effect, it is unlikely that production could begin to meet demand and permit restoration of stocks until the war was over. Further, even a short war could significantly reduce stocks of some vital munitions, and this is a matter for concern.

Distribution

Another issue closely connected with production is that of shipping supplies to the theater. In World War II, ammunition and POL comprised the greatest tonnage of supplies. Although the oilfields and refineries of Saudi Arabia and the other Gulf states ensure that POL supplies will be plentiful, the same is not necessarily true for ammunition. All the ammunition required for US Army forces, and for most of the other UN contingents, will have to be shipped in. Only local Arab forces, and US Marine Corps units operating from pre-positioned depots, will not require the creation of extensive stockpiles prior to the start of operations. Further, ammunition is not the only type of supply which must be shipped to the theater. Medical supplies, spare parts, tools, construction materials, administrative supplies, and many other kinds of material will have to be brought to the theater, and organized and assembled in depots prior to distribution. Scattered news reports from "the Gulf" in late autumn 1990 indicate that there have been some problems in this regard, not in relation to sheer quantity, but rather in the realm of supply administration and control.

Two other factors must be considered. Traditionally the US Army operates on an inventory pipeline to an overseas operation that requires 120 days from order to delivery. This 120 days breaks down as follows: there is a 30-day operational level, a 30-day safety level, 30 days in land and sea shipping time, and 30 days in the Zone of the Interior (the United States) to acquire, organize, and load the materials for shipment by the vendor and/or depot. This 120 days can be cut in half, at great expense, by air delivery—if the airlift is available.

There is another factor. During and since the Vietnam War, US forces have operated on the "Individual Line Item Requisitioning System," making use of computers, and many-digited numbers. If a cotter pin is ordered, and two digits should be reversed, a dump-truck may be delivered. In the event of an accident, a power-failure, or nearby explosion of an incoming missile, the system will be paralyzed for a longer or shorter period of time. Anyone familiar with computer "crashes" can understand that gremlins abound and glitches prevail. The famous Clausewitzian concept of "friction" applies to modern high technology not only on the battlefield, but in the support system.

Producing the required supplies and moving them to the theater of operations is only half the task, at best. The most challenging task facing supply units is actually distributing the required items to the units which need them. This process, which is relatively easy when all of a force's units stay in the same place from one day to the next, becomes much more difficult when units are moving, engaging in combat, losing equipment, and both transferring sub-units to, and receiving other sub-units from, neighboring units. All of this combines to make supply difficult, and problems such as delays in receiving required items are fairly common. If these problems are not corrected quickly they can have a serious effect on a force's ability to complete its mission.

It was in this area, the transport and distribution of supplies in-theater, that created particular problems for Rommel in North Africa. His repeated, and brilliant, offensives against the British, especially the one that carried him as far east as El Alamein, hopelessly overstretched the minimal surface transport resources available to the Afrika Korps and its Italian allies, and every mile Rommel advanced eastward from Tripoli worsened his supply situation. Although US and allied forces in the Gulf region possess much greater surface and aerial transport capacities than did Rommel in 1941 and 1942, the problem of distribution is still a major logistical concern.

In some respects distribution is a problem in the Gulf because the area is, in the modern American military phraseology, not a "mature theater." Put another way, the dense infrastructure of roads, railways, airfields, ports, buildings and other structures which are so common in the industrialized nations do not, by and large, exist among the Gulf states. In large part because their populations are fairly small in relation to the land area they cover, these countries have not developed many of these features. Some facilities, such as petroleum product pipelines, exist in comparative abundance, but these generally have not been constructed with an eye toward potential military operations and so would be of only limited utility in the event of war.

On the other hand, one logistician commented to us: "If we had to choose a war theater, we couldn't pick a better place than the Gulf Region; it is far better than any place we fought in World War II, Korea, or Vietnam." He bases this observation on the fact that Saudi Arabia, Bahrain, and Dubai boast three of the best ports in the world, with abundant deep draft berths, hundreds of thousands of square feet of covered storage, with additional air-conditioned and refrigerated warehouses. He concluded by saying: "We also have the novel advantage of going into a rich country that is ready to pay our freight. And where we don't have to transport any petroleum, oil, or lubricants. Wow!"

The facilities are of course adequate for the Royal Saudi Army, which contains only nine brigades (the equivalent of less than three US divisions). However, despite the unexpected bonuses pointed out by our logistician friend, the overall facilities are not adequate to support eight US Army divisions, two Marine Corps divisions, two Marine Expeditionary Brigades, British and French forces, and divisions from Syria and Egypt, not to mention smaller forces from the Gulf states and several African countries. This means that there is a lot of construction going on, and that any further deployment to the area compels construction of additional facilities. These include roads, airfields, hospitals, communications processing centers, and other facilities. In a sense, the US and its allies are trying to create in a matter of three or four months what has evolved in areas like West Germany over the course of the last three or four decades.

The hastily constructed depot facilities are being operated by soldiers who do not have the experience of the largely civilian specialists who operate the depots in the United States and in Germany. There is reason to believe that they have not yet developed the capability to operate their facilities as efficiently as will be required for high intensity combat operations.

One major mitigating factor in this relatively sobering logistics picture is the UN naval presence in the Persian Gulf and other nearby waters. The US Marines, for example, depend for a good deal of their logistical support on the US Navy. The Navy's amphibious assault ships contain aircraft maintenance facilities, and Navy hospital ships allow the Marines (and the Navy itself) to dispense with building a network of land-based medical facilities, such as the Army has had to do. Furthermore, naval supply ships, prepositioned in the Indian Ocean and now in the Persian Gulf, can supply the Marines with many expendable supplies for a considerable period of fighting.

Moreover, while there are some 800 to 1,000 US Air Force combat aircraft in the Gulf, there are nearly 500 Navy aircraft on the 6 carriers in the Gulf region or available nearby, and over 300 Marine Corps support aircraft. One of the great virtues of aircraft carriers, especially the larger ones, is that however expensive they may be, they are essentially self-contained, with virtually all the facilities their air wing requires contained on board. This ability to operate away from shore support for extended periods is the main reason why aircraft carriers have become the principal instrument of late 20th century "power projection." It also means that the US Navy will be able to maintain its contribution to the air campaign against Iraq without burdening either the air or land transport networks in-theater.

It is also worth considering that all of these logistical concerns affect the Iraqis too. Like the allies, their fuel supply is probably fairly safe, and they have a fuel-rationing plan available (tried out briefly in August, but withdrawn because of unpopularity). Neither are food and other subsistence supplies immediate cause for concern, and Iraq has maintained the military infrastructure created during the war with Iran, so its army is not faced with the problems of construction which confront the UN allies. However, Iraqi supplies of munitions and major end items (US classes V and VIII) are more limited than those of the allies. Iraq has not been able to import munitions since early August, and domestic production facilities, while impressive by regional standards, are limited. Particularly, the Iraqis will be unable to replace their expenditure of antitank, antiaircraft, and air-to-air missiles, and replenishment of some types of artillery and gun ammunition will be difficult.

Further, the Iraqis will be unable to replace combat losses of major equipment like tanks, other armored vehicles, artillery, and aircraft. Their ability to repair damaged equipment will also be hampered by a lack of external supplies. Although the political position of the UN allies will degenerate in the event of a long war, Iraq's armed forces will be faced with devastating supply problems should the war last longer than a month or two, and they could encounter significant shortages much sooner than that.

Summation

Logistics constrains strategy. The capacity of a force to support its operations places limits on what goals it may achieve. The US and its allies suffer from no major limitations in terms of resources, and have adequate supplies of material within the theater of operations. What could cause problems for the allies is not the simple "supply" of material resources, but rather their distribution. Adequate transport resources, and the need for air protection of overland supply routes, are crucial to allied military success. Further, in the event that allied forces have to advance against Baghdad, they could face significant local supply and support shortages, caused by the inability of the support commands to move the required materials over extended distances to the units which need them.

The Iraqis also face significant logistical problems. First, their transportation system will be under heavy aerial attack and interdiction, which will severely limit both their ability to maintain depots, and their ability to move support materials, replacement personnel, and supplies to units at the front. The effect of allied airpower will be magnified by Iraq's comparatively limited transportation system; there are few if any alternate routes available. There is also little covering terrain; there are no trees to park supply trucks underneath while enemy fighter-bombers are overhead.

Iraqi forces also will be unable to secure replenishment of expended munitions, especially complex guided weapons like antitank missiles. This means that as the war goes on Iraqi forces will be at an increasing disadvantage vis-á-vis allied forces, although this is unlikely to be a significant factor until 10 to 15 days into the war. Iraq's forces, especially those deployed in Kuwait and the southern Iraqi desert, will suffer because of interdicted supply lines. Facing UN air supremacy, Iraqi support units will be hard-pressed to deliver supplies to the front, and may well find the task nearly impossible. Even forward-deployed stockpiles may not help, since those would become major targets for UN air attack.

Logistical considerations will impose limitations on both sides' operations. The allies will have difficulty adequately supporting operations in areas where they have not been able to improve the transport infrastructure. For their part, the Iraqis will face grave problems in munitions replenishment, equipment repair and replacement, and transportation, especially in the event that war last for more than a few weeks.

CHAPTER 11

Option Five: Operation "Siege"

One of the distinguishing features of modern warfare is that it is much easier to supply units if they remain in one place for extended periods, engaged in predictable activities, than if these same units are actively campaigning and continually demanding unexpected quantities of fuel, munitions, and equipment (see Chapter 10, The Sobering Effect of Logistics). However, while static and predictable activities are beneficial logistically, they do not usually win wars. The benefit of Operation "Siege," however, is that it combines a generally undemanding logistical stance with limited ground operations and an intensive air effort in such a way as to weaken Iraqi forces with minimum cost in lives and materiel to the UN forces, while setting the stage for other operations.

Description

Operation "Siege" will be initiated approximately five to ten days after the beginning of the air campaign of Operation "Colorado Springs" (see Chapter 5). While the air campaign is continuing, UN forces of the Eastern and Western armies will undertake a campaign of probes and limited attacks against Iraqi forces in southern Kuwait and Iraq. These attacks will not try to gain geographical objectives, or even to destroy Iraqi equipment and units (although that would be a welcome side-effect), but rather would try to keep the Iraqis off-balance and to compel them to expend their limited stocks of munitions and other supplies at as great a rate as possible. Since Iraq's stock of munitions is in short supply, and spare parts for armored, mechanized, and automotive equipment are even more limited, Iraqi response to these problems will probably result in a significant reduction in Iraqi combat readiness.

Ideally, Iraq will also be faced by comparable demonstrations and probes from the main Syrian and Turkish armies along its northwestern and northern borders. Such operations, even on a very limited scale, would divert Iraqi attention from the main front, force an even greater dispersion and consumption of resources, and contribute to Iraq's perception of its own isolation. Although these operations by Syrian and Turkish forces would be quite valuable, they are not essential to the implementation of Operation "Siege," which would still be able to accomplish its objectives. It would just take longer.

An important advantage to a campaign of limited operations is that it would build the confidence and combat experience of UN forces. This would also give the commanders and staffs valuable experience in real operations. Further, such limited attacks and probes would serve to highlight Iraqi weakness. and expose these to the UN forces. Iraqi responses would give valuable insights regarding the way in which the Iraqis react to various UN offensive actions, and also regarding the speed of those reactions. Furthermore, this is likely to be frustrating for the Iraqi leadership, since they cannot be sure that any probe is not the beginning of a major offensive. Consequently, they will have to react to each UN initiative as if it were a major assault, rather than risk an insufficient response.

News reports in mid-December 1990 have indicated that US commanders in Saudi Arabia do not believe that forces arriving from the US Seventh Army will be ready to conduct active operations by 15 January, and probably not until early to mid-February. If these reports reflect real concerns of the commanders—and not disinformation intended to mislead the Iraqis—then implementation of Operation "Siege" would give the US and its UN allies valuable time to acclimatize their reinforcements and ready them for combat, while avoiding the appearance of wasting time. However, it should be rememberd that, historically, few commanders have had the luxury of complete readiness before entering combat.

It is unlikely that, given the apparent determination of Saddam Hussein and Iraq's Ba'athist leadership to resist pressure, Operation "Siege" would bring about an Iraqi capitulation. Therefore, after a period of ten days to two weeks or more, the UN forces would have to mount a ground offensive to expel the Iraqis from Kuwait and inflict on their armed forces a defeat of sufficient magnitude to compel them to end the war. The potential of success for such an operation, in the wake of both air and ground efforts made to reduce the Iraqi capacity for resistance, would have increased substantially, and a major offensive along the lines of either Operation "Leavenworth" (Chapter 7) or the more elaborate and somewhat riskier "RazzleDazzle" (Chapter 8), would produce significant results within a short time. In such a case, losses for these follow-on ground operations would almost certainly be lower than if they had been launched shortly after the war began.

Problems with "Siege"

The principal drawback in besieging Iraq in this fashion is that it will take time. This in itself is not particularly serious, although it would be best if all major combat operations were concluded before either the April "windy season" or the May heat arrive. Therefore, to conclude operations by mid-March, and allowing for an overall campaign of 40 to 60 days, it would be necessary to begin major hostilities (with "Colorado Springs") shortly after the expiration of the UN ultimatum on 15 January, 1991.

The more pressing time-related problem is not military, but rather political. Opposition to national policy in the Persian Gulf area among the public of several western nations, notably the US, Great Britain, and France has so far been comparatively muted. Furthermore, the commencement of hostilities would likely produce a patriotically-inspired surge in public support for the war, especially in the US. Public fears of a long war, (again, especially in the US) scarcely would be assuaged, however, by the implementation of low-keyed Operation "Siege." The two- to four-week period between the commencement of air operations and the beginning of a major ground offensive might result in a significant erosion of public support for the war, and this could have serious political consequences.

If US forces not fully ready to begin operations until February, the gap between 15 January, 1991 and the actual commencement of operations could be a serious political liability, and could adversely affect the morale of UN troops in the area. Thus, implementation of Operation "Siege" would, at the very least, show that the UN forces are doing something, and such a continual program of limited attacks and probes would help maintain morale, thereby reducing the effects of having to wait until the whole force is ready—presumably sometime in the first half of February. Regardless of sound military reasons, postponement of offensive operations until February would be politically damaging.

Political leaders in the western democracies (again, especially those in the US) therefore would have to make a difficult choice concerning Operation "Siege." While its implementation would yield valuable military benefits, the domestic political cost associated with it may be too high to offset the relatively small number of killed and wounded servicemen returning home. Put another way, a one-month war with relatively high casualties may be more politically acceptable to the public than a two-month war with lower casualties.

Effect of Iraqi Responses

Essentially, Iraq will have four main strategic choices should war begin against the UN forces in the Gulf region. First, Iraq may launch a pre-emptive spoiling attack against either the US, British, and French forces along the Gulf Coast or against the Arab forces inland to the west. Second, Iraq could invade Jordan, endeavoring to seize Jordanian airfields and capture Amman. Successful execution of this course of action would expand the conflict geographically, raise the political 'stakes,' and would probably draw Israel into active participation. Third, Iraqi forces could opt to hold their current positions and "slug it out" in southern Iraq and Kuwait. Finally, the Iraqis could withdraw from all but the northeastern portions of Kuwait (Bubiyan Island and the Rumaila oil field), leaving most of the expected UN offensive plan to strike thin air.

As far as Operation "Siege" is concerned, an Iraqi spoiling attack would have little effect. The bulk of UN forces not committed to probes of their own would be in defensive positions, or would be getting ready. Further, the Iraqis would be leaving the advantage of their defensive positions and encountering UN defenses. Iraqi forces have shown little capability to conduct offensive operations in any of the four conflicts in which they have been engaged over the past half-century. Finally, any Iraqi offensive action would expend valuable munitions, and thereby contribute to one of the goals of Operation "Siege," namely the reduction of Iraqi munitions, spare parts, and general supply stocks. In a real sense, any Iraqi offensive action against UN forces in Saudi Arabia would contribute to achieving the objectives of Operation "Siege."

The possibility of an Iraqi attack on Jordan has been discussed earlier (Chapter 9).

An Iraqi withdrawal to the extreme northeast of Kuwait would make a good deal of sense militarily. Such a move would seriously reduce the effectiveness of any of the UN forces' basic offensive options, and would drastically shorten the Iraqi supply line. Despite repeated Iraqi declarations that Kuwait is now the nineteenth province of Iraq, it is possible that Saddam Hussein will order his forces to withdraw from their newest province before a UN attack. This potential "January surprise" is discussed in Chapter 13.

All in all, there is not much that Iraq can do to frustrate Operation "Siege." All that the Iraqis can hope for is that they repulse the probing UN forces so effectively that the war becomes politically unpopular in the western democracies.

Assessment

Operation "Siege" is, in military terms, perhaps the most sensible of the basic courses of action available to the UN forces, and entails the fewest military risks. It promises limited casualties for the first two or more weeks of war, and in the event that a major offensive is later unleashed, the Iraqi defenders will have been significantly weakened by air attacks and ground probes. Additionally, the implementation of Operation "Siege" will provide the UN forces with a valuable period in which to acquaint themselves with the rigors of desert warfare, and to gain combat experience. This could be particularly important, since only a small fraction of the UN forces has been in combat as units within the past decade. Carefully planned attacks, orchestrated in such a way as virtually to ensure success, would do much to make the troops taking part more confident, while at the same time exposing them and their commanders to the realties of combat situations.

On the other hand, the two to four week time period involved in implementing Operation "Siege" would be a significant political liability. The impression of stalemate which would almost unavoidably be connected with Operation "Siege" by the public in the US and other western countries would do little to enhance support for the war or for UN policy in the Gulf. The longer a war goes on (and implementation of Operation "Siege" could mean a somewhat longer war), the more likely that the fragile UN coalition against Iraq would begin to unravel.

CHAPTER 12

The Decision

The procedure for carrying out an Estimate of the Situation, as shown in Figure 6, Chapter 4, provides first for comparing all of the courses of action under consideration with all of the enemy's capabilities as assessed by our own intelligence organization. In the United States Army this is usually a cooperative process, between the G-2 (the senior Intelligence staff officer) and the G-3 (the senior Plans and Operations staff officer) or—more likely—members of their respective staffs. This is—or should be—a painstaking process, which then is followed by another painstaking process in which the G-3 and his staff make a final comparison of the courses of action. For our purposes we can assume that all of the previous analyses and comparisons have been made, and we shall review only the results of the final comparison, and the decision which follows.

The Alternative Courses of Action

It will be remembered from Chapter 4 that the five principal courses of action are:
- Operation "Colorado Springs." This is a two-phased air campaign which in its first phase is directed against the general war-making capability of Iraq. The second phase is to be directed against the Iraqi ground forces deployed in and near Kuwait. This is not really an alternative course of action, but rather is seen as a preliminary to

all of the four possible alternative ground courses of action. This course of action might conceivably cause so much damage to the Iraqi military establishment as to bring about a surrender, thus obviating the need for any of the ground courses of action. This, however, is far from certain; a ground army has never before been defeated by the application of airpower only.
- Operation "Bulldozer." This is a classic "attrition warfare" scenario, a frontal attack against the fortified positions of the Iraqi South Kuwait Defensive Force, using firepower and brute force (which are available in abundance) to overwhelm the defenders.
- Operation "Leavenworth." This scenario, also directed against the South Kuwait Defensive Force, is intended to accomplish the same purpose as "Bulldozer," but to put more emphasis on maneuver by carrying out a double-envelopment, using armor as the maneuvering force on the left (west), and two Marine Expeditionary Brigades (MEBs) as the amphibious spearhead of the maneuvering force on the right (east).
- Operation "RazzleDazzle." This is similar in concept to "Leavenworth," but the envelopments are deeper. It envisages two wide turning movements (or strategic envelopments), combined with an air assault, to encircle all Iraqi forces in Kuwait.
- Operation "Siege." This scenario provides for a two-phased ground operation. The first phase will be carried out simultaneously with an open-ended continuation of the second phase of "Colorado Springs." The principal activity in this first phase will be a series of relatively small-scale probes carried out along the fronts of both the Eastern and Western armies against the positions of both the South Kuwait Defensive Force and the South Iraq Defensive Force. No major ground offensive will be undertaken during this phase, which will last for a week or more. Syrian and Turkish forces, along Iraq's northwestern and northern frontiers, will be urged to carry out similar probes of Iraqi defensive positions on those frontiers. The second phase will be one of the other three ground options ("Bulldozer," "Leavenworth," or "RazzleDazzle"), as seems most appropriate at the time.

Special Case Courses of Action

Two other, special case, courses of action are considered. Neither of these is thought of as an alternative course of action, but rather would be put into effect under a given set of circumstances, in conjunction with one or more of the five courses of action briefly described above. These two special cases are:

- Operation "Jordan." This would be a response to an Iraqi move—prior to the initiation of hostilities, or simultaneously with an Iraqi initiative—into Jordan (intended presumably to provoke Israeli involvement). This would trigger three responses from the UN Forces:
    (1) Immediate air assault or air landing (with or without Jordanian cooperation) to occupy the Jordanian air bases at Mafraq and Azraq;
    (2) Immediate air attack upon Iraqi forces entering Jordan, and along the Mafraq-Baghdad road as far east as the Euphrates River;
    (3) Initiation of Operation "Colorado Springs."
- Operation "Salah-al-Din." This is a quick move to occupy Kuwait City by a three-battalion Saudi Arabian-Kuwaiti-US task force, in the event of a sudden Iraqi collapse in southern and central Kuwait. This could be mounted as the concluding phase of any of the five basic courses of action, with back-up support provided to Task Force Salah-al-Din by US or Anglo-French forces, depending on the circumstances.

The Assessment Process

We shall assume that in its final analysis the G-3 staff will make use of a relatively simple, quick, computerized combat simulation which will forecast the results of the battles that are expected to occur for each of the ground combat courses of action. (One such simulation is the Tactical Numerical Deterministic Model, or TNDM.) In running these simulations the staff officers will consider two of three possible situations with respect to Operation "Colorado Springs": the effectiveness of the ground defenders has not been seriously impaired, or their effectiveness *has* been seriously impaired. (The third possibility, which obviates the need for use of a ground combat simulation, would be that the defenders have been rendered ineffective by the air campaign, and have either collapsed or surrendered.)

Using a simplified methodology based on the TNDM model it is possible to calculate the relative combat power of the forces engaged, and to calculate the casualties on a daily basis. Without more detailed knowledge of the strengths, organizations, and actual deployments of the forces, it is not possible to make precise calculations. Anyone who believes that it is possible to make precise calculations of the future is deceiving himself, or his audience, or both. However, the data available to us is comprehensive enough to allow us make quite reasonable approximate projections which—on the basis of similar estimates of historical experience—are probably accurate to within plus-or-minus 50 percent. (This methodology is discussed in some detail in: T.N. Dupuy, *Attrition: Forecasting Battle Casualties and Equipment Losses in Modern War*, HERO Books, McLean, Va., 1990, chap 7.)

This methodology is applicable to all of the major courses of action except Colorado Springs which—by itself—involves no ground actions. Five assumptions will provide a basis for estimating the aircraft losses and casualties to be expected among the UN aerial participants in this operation. The assumptions are:

1. There will be 3,000 UN sorties on the first day (most by US aircraft), and approximately 2,000 sorties per day for the next nine days, or 21,000 sorties over the ten days.
2. The loss rate will be two percent the first day, with the rate dwindling to about 0.75 percent on the tenth day, for an average loss rate of 1.21 percent for the ten days, or a total loss of 255 aircraft in those ten days.
3. Somewhat conservatively it is estimated that there will be 380 American casualties among the airmen, of whom half will be killed and half wounded.
4. If, at the end of ten days, Iraq has not surrendered, then ground operations will begin.
5. After ground operations begin, air operations will continue at the same sortie rate, with ten casualties per day for the next 20 days, and five casualties per day for the next ten days.

The results of these assumptions are shown in Figure 19.

Also shown on Figure 19 are the results of the casualty calculations for:
  a. Operation "Bulldozer," the ground operations alone, and the total, including casualties from the air campaign.

Operation	D-Day to D+9			D+10 to D+19			D+20 to D+29			D+30 to D+39		
	Dead	Wound	Total	Dead	Wound	Total	Dead	Wound	Total	Dead	Wound	Total
Colorado Springs	190	190	380	50	50	100	50	50	100	25	25	50
Cumulative				240	240	480	290	290	580	315	315	630
Bulldozer				1618	8092	9710	160	840	1000			
Cumulative							1778	8932	10710			
With Air				1858	8332	10190	2068	9222	11290			
Plus 50%				2428	12137	14565	240	1260	1500			
Cumulative							2668	13397	16065			
With Air				2668	12377	15045	2958	13687	16645			
Leavenworth				1214	6069	7283	160	840	1000			
Cumulative							1374	6909	8283			
With Air				1454	6309	7763	1664	7199	8863			
Plus 50%				1821	9094	10915	240	1260	1500			
Cumulative							2061	10354	12415			
With Air				2061	9334	11395	2351	10644	12995			
RazzleDazzle				1079	5394	6473	160	840	1000			
Cumulative							1239	6234	7473			
With Air				1319	5534	6853	1529	6524	8053			
Plus 50%				1619	8091	9710	240	1260	1500			
Cumulative							1859	9351	11210			
With Air				1859	8331	10190	2149	9641	11790			
Siege				324	1618	1942	540	2697	3237	160	840	1000
Cumulative							864	4315	5179	1024	5155	6179
With Air				564	1858	2422	1154	4605	5759	1339	5470	6809
Plus 50%				726	2667	3393	860	4096	4956	240	1260	1500
Cumulative							1586	6763	8349	1826	8023	9849
With Air				966	2907	3873	1876	7053	8929	2141	8338	10479

FIGURE 19.
UNITED STATES FORCES' CASUALTIES FROM D-DAY TO D+40

b.	Operation "Bulldozer," allowing for a possible 50 percent underestimation, also with air casualties added.
c.	Operation "Leavenworth," the ground operations alone, and the total including air casualties.
d.	Operation "Leavenworth," allowing for a possible 50 percent underestimation, also with air casualties added.
e.	Operation "RazzleDazzle," the ground operations alone, and the total including air casualties.
f.	Operation "RazzleDazzle," allowing for a possible 50 percent underestimation, also with air casualties added.
g.	Operation "Siege," followed by Operation "RazzleDazzle," the ground casualties alone, and the total including air casualties.
h.	Operation "Siege," followed by Operation "RazzleDazzle," allowing for a 50 percent underestimation, also with air casualties added.

For each of the above ground operations, an additional ten days is estimated for an advance to occupy Baghdad, and figures are included for the ground casualties alone, and the air casualties added, as well as the cumulative losses for each of these operations.

The Cost to Iraq

Figure 20 shows the cost to Iraq of such a campaign. Iraqi casualties under the punishment of Operation "Colorado Springs" are estimated at about 30,000, of whom 6,000 would be killed and 24,000 wounded. Regardless of the UN option, the Iraqi casualties during the next ten days of operations on the southern front would be about 35,000, with 7,000 killed and 28,000 wounded. Losses in the third ten-day period would be about 38,500, with 7,700 killed, and 30,800 wounded. If the war should last another ten days, with the Iraqi forces in a state of collapse, casualties would decline to about 15,000, with 3,000 killed and 12,000 wounded. The cumulative figures show the total losses in killed and wounded if the Iraqis surrender after 10 days, after 20 days, after 30 days, and the total losses of the overwhelmed Iraqi army if the war lasts 40 days, by which time the Iraqi Army would have disintegrated. It will be noted that the proportion of killed to wounded is different for US forces and Iraqi forces. Both are based upon historical relationships. The difference is that it is assumed that US troops will be wearing flak jackets, and that the Iraqis will not.

Operation	D-Day to D+9			D+10 to D+19			D+20 to D+29			D+30 to D+39		
	Dead	Wound	Total	Dead	Wound	Total	Dead	Wound	Total	Dead	Wound	Total
Colorado Springs	6000	24000	30000									
Other Ground Operations				7000	28000	35000	7700	30800	38500	3000	12000	15000
Cumulative				13000	52000	65000	20700	82800	103500	23700	94800	118500

FIGURE 20. IRAQI CASUALTIES

Operation	D-Day to D+9			D+10 to D+19			D+20 to D+29			D+30 to D+39		
	Dead	Wound	Total	Dead	Wound	Total	Dead	Wound	Total	Dead	Wound	Total
Colorado Springs	38	38	76	10	10	20	10	10	20	5	5	10
Cumulative				48	48	96	58	58	116	63	63	126
Bulldozer				486	1942	2428	50	200	250			
Cumulative							536	2142	2678			
With Air				534	1990	2524	594	2200	2794			
Plus 50%				729	2913	3642	75	300	375			
Cumulative							804	3213	4017			
With Air				777	2961	3738	862	3271	4133			
(Same for Leavenworth & RazzleDazzle												
Siege				97	386	483	162	647	809	50	200	250
Cumulative							259	1033	1292	309	1233	1542
With Air				145	434	579	317	1091	1408	626	2324	2950
Plus 50%				146	579	725	243	971	1214	75	300	375
Cumulative							389	1550	1939	698	2783	3481
With Air				194	627	821	447	1608	2055	761	2846	3607

FIGURE 21. UN ALLIES' CASUALTIES FROM D-DAY TO D+40

The Cost to the UN Allies

For a variety of reasons, discussed to some extent in Chapters 3 and 4, United States ground forces will bear a much heavier combat burden than will the UN allies. Whether or not this is right and equitable is debatable. However, it seems unavoidable. It is likely, however, that (commensurate with the different sizes of contingents), air contingents of the UN allies will incur casualties at roughly the same rate as the US air forces. Figure 21 shows the casualties anticipated the UN allies over the same periods as for US and Iraqi forces.

Decision

If the above figures are close to accurate (and history tells us they should be), then the proper solution is to begin the war with the air campaign of Operation "Colorado Springs." If this should result in an Iraqi surrender, so much the better. If not, then after about ten days of "Colorado Springs," to continue the air campaign for about ten more days while initiating Operation "Siege." If this should not bring about an Iraqi surrender, the ground campaign should be concluded with Operation "RazzleDazzle." If this has not brought about an Iraqi surrender, then an advance should be made through the desert to destroy any resisting Iraqi forces and to occupy Baghdad if necessary.

CHAPTER 13

After the War is Over

Any effort to predict "the shape of things to come" is necessarily fraught with danger. This is more particularly the case when speaking of the future in one of the world's most volatile regions. However, assuming that the UN allies pursue the war to a successful conclusion as outlined in the preceding chapters, the immediate post-war situation in the theater is, relatively, predictable. Regional prospects for the mid- and long-term may be sketched in broad outline, but are neither as apparent nor as predictable.

"January Surprise" Possibility

What follows is predicated, like this book, on the likelihood of war. There is of course the possibility of a "January surprise" in which the wily Saddam fools everyone (or in hindsight no one) by "backing down" on Kuwait and offering to get out of Kuwait altogether or compromise on details of the land grab—say returning the bulk of Kuwait but retaining an oil field and some islands in the northeast. Although this would still appear unlikely in view of the scale of the Iraqi buildup in Kuwait and adjacent areas, it cannot be ruled out.

If such a situation transpires, it may be tempting to "declare victory" and accept a compromise. That however would simply reward the aggressor and would do nothing to implement UN Security Council resolutions. Nor would it prevent a recurrence of the aggression in the future. It would, in fact, be likely to ensure future Iraqi aggression on a much greater scale.

The appropriate response to a "January surprise" would be to reject outright any Iraqi compromise offer, insist on Iraqi acceptance of all UN Security Council resolutions, and demand that the Iraqi army begin to evacuate Kuwait immediately. Perhaps a 48-hour extension of the deadline might be offered, as suggested at the end of Chapter 2. But no more. This period would be designated as a period of "pre-hostilities," during which the allies would themselves immediately initiate military activities preparatory to moving into Kuwait and engaging any remaining Iraqi forces.

Assuming that Saddam would accept this renewed ultimatum, the situation at that point would be like that obtaining in 1918 when, after the failure of German arms and the declaration of the Armistice, the German armies on the Western Front were permitted to make an unmolested "homeward march." The result for Iraq, of course, would be disarmament and a dictated peace, but the country would be spared the devastation attendant on military operations.

Immediate Post-War Situation

The remainder of this chapter assumes that war will break out, and will follow a course such as is suggested in previous chapters. Besides the strictly military problems and preoccupations that occur as a consequence of any battle or campaign (for instance, care of the wounded, and minefield clearance), there will be a complex civil affairs task, possibly greater than any such job since the Korean Conflict. Among other things, this will involve the relationship of allied forces with civilians and the reconstruction of Kuwait and any areas of Iraq or other nations devastated by Iraqi occupation and by fighting. In wartime this is a job carried out principally by military civil affairs specialists (G-5) with the cooperation and assistance of civil authorities. In a post-war situation, the military role diminishes over time as conditions return to normal and the civil authorities gradually reassume all their functions.

Kuwait

Perhaps the most pressing immediate problem will be the succor and reconstruction of Kuwait, "Saddamized"—as one US Congressman put it—by one of the most cruel aggressions of the century. It would appear, from news reports, that Kuwait's infrastructure has been systematically dismantled and its people brutalized and, to an extent, displaced by Iraqi soldiers and "settlers." The complicity of Kuwait's large Palestinian population in the pervasive terror of the occupation has been documented, but it is by no means clear that Palestinian collaboration has been widespread.

Iraq

It is most unlikely that Saddam Hussein's regime, which is coextensive with the Iraqi Ba'ath Party, will survive the war. Many view this possibility with concern, fearing that a political vacuum would result, producing a dangerous instability in Iraq and in the region. They argue that in this event, the UN-administered "medicine" could be worse than the malady it sought to cure.

It is difficult to see how this argument can be offered seriously. There has recently been a dangerous instability in the Middle East due to the overwhelming (regionally) military power built up by Saddam. This was demonstrated by the invasion of Kuwait and Saddam's very real threat to Saudi Arabia, as well as his continual threats to strike at Israel with all means he possesses, including implicitly weapons of mass destruction. It may be useful at this juncture to ruminate upon the strategic and tactical/operational situations would be *if* Israel had not destroyed Iraq's Osirak nuclear plant in 1981.

If Saddam's regime is destroyed, Iraq will be occupied by the UN allies. The peace settlement will undoubtedly see to it that little opportunity is given to Iran or to Assad's Syria to take over the destabilizing superiority that Saddam enjoyed. Certainly the power of both Iran and Syria will be enhanced at the expense of Iraq. However, this will be a move back toward balance, not toward greater instability.

Saddam: War Criminal

"Vae victus" (woe to the vanquished) is a phrase normally associated with the Roman amphitheater, but in recent times it has become applicable to those whose crimes against world peace and security and the laws of war have branded them as war criminals. Capital punishment or imprisonment, such as meted out at the Nuremberg and Tokyo trials of Axis war criminals following World War II (1945-1946), should be the lot of Saddam and his henchmen, if they are convicted.

Evidence is accumulating that Saddam Hussein and various of his Ba'athist confederates, civilian and military, may be war criminals under definitions established by international law. Certainly, unprovoked aggression against a sovereign (and virtually defenseless) state, coupled with its wanton, merciless dismantlement and ruin by occupation forces, have provided grounds enough for indictment. Further, it is clear that Iraqi officers and soldiers, obviously with official encouragement, have committed innumerable violations of the laws and customs of war against the civilian population in Kuwait, to include: murder, individual and collective; mass deportations; torture; and inhumane treatment. These crimes have been documented by Kuwaiti and Western refugees and by international organizations like Amnesty International. As expected, and in the face of international revulsion, Iraq denies these allegations.

Ante-up for Terrorists

Baghdad's support for the Palestinian cause has made it a major base of international terrorist organizations, including the infamous Palestine Liberation Front—the group responsible for the *Achille Lauro* hijacking (7 October, 1985). The Front's leader, Abu Abbas, has made the threat that if the UN allies attack Iraq, terrorist revenge attacks will be launched by "striking Palestine groups" against "American and Western targets." He specified: "Among those targets are US installations and interests spread all over the globe" (as quoted in the Syrian government's *Al-Jumhuriya* newspaper, 17 December, 1990).

Abu Abbas' statement was not unexpected, and its sentiment has been more or less repeated several times since by other terrorist spokesmen, including representatives of Yasser Arafat's PLO. These terrorists are much indebted to Saddam Hussein, who has provided them a secure sanctuary and consistent financial support in the past few years. However, the allied offensive, when it comes, will remove both sanctuary and financial backing. It is not too much to suppose that terrorists like Abu Abbas (who was once nearly captured by US troops) and terrorist organizations and infrastructures are among the secondary objectives of planned military operations.

Certainly the terrorists themselves know this, and they are undoubtedly feeling vulnerable. The broad base of the allied coalition and the fact that Iraq is *nearly* surrounded by unfriendly nations (Iran's benignancy toward

Saddam is suspect, and Jordan will not harbor terrorists) leaves them exposed as they never have been before.

The number of sanctuaries is diminishing, and for the particular groups clustered under Saddam's roof, Damascus is no longer an alternative. That leaves Lebanon—at least those areas not subject to Israeli or Syrian control—and Libya, with perhaps Sudan and Yemen as unattractive last resorts. The problem for the terrorists, however, may lie in getting to any of these places in the first instance and, having gotten there, surviving without the level of financial support ("state sponsorship") enjoyed in more halcyon days.

The potential of terrorist organizations to wreak havoc against "American and western targets" is undoubted, but security organizations worldwide will be on full alert when the balloon goes up and actual terrorist capabilities will be greatly diminished. On the other hand, allied capabilities to damage and destroy terrorist organizations by military operations, including special operations, will be stressed in pre-operational planning, and if terrorists show themselves, or are flushed, they may not expect to be treated with the same consideration they may formerly have come to expect from certain western and non-aligned governments anxious to speed them on their way "home."

A New Strategic Alliance

Coalitions are notoriously fragile enterprises, and where they have succeeded it has been more often than not because of the emergence of strong leaders whose vision transcends mundane considerations of purely national interest. These leaders have to articulate that vision for coalition members in clear, straightforward, easily understood terms, so that there is no room for disagreement as to goals and objectives.

The US-led, UN-sanctioned coalition, forged with firm purpose by President Bush and guided through a veritable maze of parochial, disintegrative pressures by the president, would not appear to be in danger of dissolving at the successful conclusion of the war. There is simply too much at stake—too much invested in men, treasure, and materiel (not to mention individual political futures)—for the coalition's leadership to permit it to collapse following the transitory euphoria of military victory.

One guarantor of the coalition's continuing effectiveness is that its leadership is irrevocably vested in its architect, President Bush. Some of history's more effective coalitions have essentially been partnerships, either at the political level (Roosevelt and Churchill) or the military level (Marlborough and Eugene; Washington and Rochambeau). It is fair to observe, however, that although several of the Arab leaders, most notably Egypt's Mubarak, have displayed great courage, the one coalition leader who might have deserved equal billing with the president—Britain's Margaret Thatcher—has left the world stage. Bush will likely emerge from a victorious campaign as a world hero, and will be much strengthened politically at home (in what has been a disappointing year for him domestically).

Bush's strength abroad and at home, and the prestige of military victory and success at the world political level, will help to sustain the coalition in the difficult transition to regional peace. At the same time stability will provide him with the opportunity to forge a new post-Cold War strategic coalition based on the UN-sponsored alliance.

Some commentators have predicted post-war chaos in the Moslem world, particularly if the US invades Iraq. These analysts see moderate Arab states like Saudi Arabia and Jordan destabilized in a wave of Moslem indignation that will end with the US "booted out" of the Middle East and with Syria (or perhaps currently enervated Iran) emerging as the dominant regional player. This is a scenario described not only by some American journalists but also touted (in rather more generalized terms) by Zeid Wehbe, spokesman for PLO Chairman Yasser Arafat, who warned Bush that he would have to deal with "armies you don't even know," "a broad Arab-Islamic current" that would include Iran and the Egyptian people. (It is worth noting that at least one US journalist fears that the Soviet Moslem minorities will join in the general Islamic *jacquerie*—though what bearing this would have on the course of events in the Middle East is not explained.)

This scenario ignores some plain facts and some equally transparent trends.

The plainest fact is that Saddam is militarily isolated and politically very nearly so. His "allies," a quite lukewarm lot, *may* include Libya, Sudan, Yemen, Cuba, and Jordan. None of these states will fight *for* Saddam, and one—Jordan—may itself become a victim of Iraqi aggression. Some "allies" that likely will fight for Saddam are Iraqi-sponsored terrorist organizations (discussed above) and Palestinian volunteers (most displaced or thrown out of work or subsidy by the very aggression they condone). It is doubtful that Saddam's military or political capital is increased to any noticeable degree by these accretions.

On the other hand, with those exceptions, world public opinion and Arab opinion, to the extent it can be gauged, is firmly set against the rapacious aggressor with Arab blood on his hands. If that were not so, there would be no condemnation of Saddam's actions by world and regional organizations and by Islamic leaders, no universal revulsion at the aggression of one Arab state against another, followed by a dismantling reminiscent of Nazi attempts to erase villages like Lidice in Czechoslovakia during World War II.

To be sure, there is a potential for a Moslem *jacquerie*, but that potential has always been there and has not been realized—not in August 1990 or since, not in 1982, and not in 1973, 1967, 1956, and so on. It did not even happen in May 1941 when, as we have seen, a popular revolt combined with resolute military action would have swept the British from the Middle East. Viewed in historical perspective, the likelihood of a mass, anti-Western movement in the Middle East is remote.

Another consideration, particularly relevant when discussing the potential destabilization of the moderate Arab states is that these states have inevitably faced attempted destabilization several times in the recent past and have in each instance managed to deal with the attempts firmly and decisively. One would expect that should attempts at destabilization be made in connection with the present crisis, the Arab leaders will not be surprised and will be prepared to respond with appropriate action.

Finally, with reference to dire, Cassandra-like predictions of the region rising in popular fury to expel the coalition forces, it might be well to reflect for a moment on what the situation might have been had Saddam not been checked at the Saudi frontier, and had he managed to complete his aggression by becoming hegemon of the Gulf Region, master of 50 percent of the world's oil, and leader of a militarily-enslaved Arab "nation."

In that case the very states now threatened with imminent destabilization—Saudi Arabia, Jordan, the Gulf emirates, and possibly Egypt—would have fallen in succession to Iraqi *anschluss*. It is not to be doubted that in each instance Saddam would have found contemporary Arab equivalents of Dollfuss and Schuschnigg, with the same dismal results for the Arab people. It will not be destabilizing if the anti-establishment local demagogues are deprived of their potential leader.

Regional Prospects

Successful conclusion of the war and emergence of a strong, enduring allied coalition will produce the preconditions for a comprehensive, lasting regional settlement. The Iraqi aggression after all is only the latest (and in many ways most aberrant) of the region's major problems. Post-war occupation and reconstruction of Iraq, will not address the most persistent and most pressing of the region's problems: the ongoing Arab-Israeli conflict and related problems of Palestinian nationalism, the *intifada*, and the "occupied territories."

As discussed above, it is neither possible nor desirable to link talks on Kuwait with discussion of the totally unrelated Palestinian issue (as Saddam somewhat disingenuously desires but which the US has categorically rejected). It was not necessary for Saddam to invade Kuwait in order to promote talks on the Palestinian question. However, with Saddam and his aggressive regime removed, the time would be appropriate for a comprehensive Middle East Peace Conference that would address the region's enduring problems and guarantee Israel's security.

Is Israeli cooperation in such a conference likely? Consider: the Israelis have said that they are prepared to address the Palestinian issue by "political means" provided there is an end to endemic violence in the occupied territories and in Israel proper. It may be expected that in the wake of a decisive coalition victory, Palestinian extremists will recognize that the *intifada*--now turning on itself with grotesque violence directed against "collaborators"—has served its purpose and that it is time to move toward the political settlement all sides desire. Such a settlement would, of course, include the status of the occupied territories and Israel's security.

Consider also that there has been informed speculation that the Israelis have been quietly negotiating with arch-enemy Syria over the future of the Golan Heights. Neither side would acknowledge the negotiation if asked, but the evidence is strong.

Speculation on the details of a comprehensive Middle East peace is beyond the scope of this work. It is enough to point out that the possibility of such a general settlement would be greatly facilitated by an unequivocal allied victory and the moral imperative such victory that would confer on the coalition. The UN, having defeated an aggressor, could assume the peacemaking and peacekeeping roles expected of it in the halcyon days immediately after its birth. Armed with that authority, the United Nations might indeed play midwife to a long-desired "peace in our time."

APPENDIX A

Chronology: Iraq and the Gulf Region, 1514 — 1991

1514-1555. Intermittent Turkish-Persian hostilities in eastern Anatolia, Azerbaijan, Kurdistan, and Mesopotamia, ending with the Treaty of Amasia (Amasyia) confirming Turkish possession of Mesopotamia.

1602-1638. Renewed intermittent Turkish-Persian hostilities, ending with final Turkish capture of Baghdad.

1725-1746. Ottoman wars with Persia; Turkey retains Mesopotamia.

1756. Kuwait's autonomous status established with the appointment of a member of the Sabah family as sheikh.

1898. Britain protects Kuwait against threatened Turkish occupation.

1899. Sheikh Mubarak ibn Sabah places Kuwait under British protection.

1902, January. Abd-al Aziz Al-Saud (Ibn Saud) and a small band of Wahhabi bedouin followers capture Riyadh in east central Saudi Arabia, initiating the struggle of the House of Saud to unify the peninsula.

1913. Turkey recognizes Kuwait's autonomy.

1914-1918. World War I. Turkey enters the war on the side of the Central Powers (October 1914); Allied campaigns on the Turkish fronts, 1914-1918. (The Ottoman Empire in the Middle East is dismantled by the post-war Treaties of Sèvres and Lausanne, 1920, 1923: Syria to French mandate; Palestine and Mesopotamia [Iraq] to British mandate.)

1919. Wahhabi attacks against Kuwait repulsed by British aircraft.

1921. Treaty of Mohammerah fixes boundary between Kuwait and Nejd (later Saudi Arabia).

1922. Treaty of al-Uqayr between Kuwait and Nejd; creation of the southern Neutral Zone.

1923. Kuwait-Iraq frontier settled.

1923, 29 October. Republic of Turkey proclaimed by Mustafa Kemal.

1926, 8 January. Ibn Saud is proclaimed King of Hejaz.

1927-1928. Renewed Wahhabi attacks on Kuwait.

1930. Anglo-Iraqi Treaty of Alliance and Mutual Support.

1932, 18 September. Ibn Saud proclaims the Kingdom of Saudi Arabia.

1934. Kuwait grants concession to Kuwait Oil Company, a joint Anglo-American venture.

1938. Oil discovered in Kuwait.

1939-1945. World War II.

1941, 2-31 May. Anglo-Iraqi Conflict. Axis-inspired revolt of Iraqi Premier Rashid Ali threatens British bases in Iraq. After brief initial success against weak British forces, the Iraqis are repulsed (Siege of Habbaniya, 2-6 May). Promised Axis aid is too little, too late. British forces, reinforced, defeat the Iraqis at Falluja (19 May), and occupy Baghdad (31 May). An armistice ends hostilities.

1941, 25 August. British and Soviet forces enter Iran to end Axis influence there and insure that the Persian Corridor for aid to the USSR will be unobstructed. Iranian resistance is minimal.

1941, 16 September. The Shah of Persia abdicates in favor of his son.

1945. Formation of the Arab League (League of Arab States). Members initially included Egypt, Iraq, Jordan, Lebanon, Saudi Arabia, Syria, and Yemen.

1946. Oil first exported from Kuwait.

1946, 17 April. Syria declares its independence from French mandate.

1946, May. Hashemite Kingdom of Jordan proclaimed in eastern portion of the British Palestine mandate.

1947, November. UN partition of Palestine (western portion of British mandate) between Arab and Jewish inhabitants.

1948, 14 May. State of Israel proclaimed.

1948, 14 May-1949, 7 January. First Arab-Israeli War (Israeli War of Independence).

1950, 29 January. Abdullah as-Salim as-Sabah becomes Sheikh of Kuwait.

1950, 17 June. Collective security treaty of the Arab League signed.

1951, 20 July. King Abdullah of Jordan assassinated at Amman.

1952, 22 July. Egyptian Free Officers' Revolution deposes King Farouk; Gamal Abdel Nasser assumes a key role in the government of General Mohammed Naguib.

1952-1955. Buraimi Oasis dispute between Oman and Saudi Arabia.

1953, 2 May. Coronation of King Hussein of Jordan.

1953, 19 August. Shah Mohammed Reza Pahlavi and Iranian royalists regain control of the Iranian government by force, ousting the regime of Premier Mohammed Mossadegh.

1955, 21 November. Baghdad Pact signed, creating US-sponsored Middle East regional security organization (the US was not a member). Members included: Iraq, Turkey, Iran, Pakistan, and Great Britain.

1956, 23 June. Nasser elected president of Egypt.

1956, 29 October-5 November. Second Arab-Israeli War and Anglo-French invasion of the Suez area (Suez crisis).

1958, 1 February. United Arab Republic (UAR) formed by the union of Syria and Egypt; prestige of Egyptian President Nasser soars.

1958, 14 July. Iraqi Revolution. Free Officers' movement led by Gen. Abdul Karim Kassem overthrows the regime of Premier Nur al-Said in a bloody coup. King Farouk II and Nur al-Said slain.

1960, 17 September-1966, 29 June. Kurdish rebellion in Iraq; settled by grant of local autonomy to Kurd rebels.

1961, 19 June. Britain grants Kuwait independence.

1961, 25 June. Iraq claims Kuwait and threatens invasion. British troops are landed in Kuwait (27 June) to protect the emirate.

1961, 20 July. Kuwait admitted to the Arab League and protected by a League peacekeeping force, forestalling Iraqi annexation.

1961, 28 September. Syria secedes from the UAR following a military coup.

1963, 8 March. Arab Socialist Resurrection (Ba'ath) Party headed by Amin Hafiz attains power in Syria following coup.

1963, 14 May. Kuwait admitted to the United Nations.

1963, October. Iraq recognizes Kuwait's independence.

1965, 1 January. Arab League creates Arab common market (thus far only Iraq, Jordan, Syria, and Egypt are members).

1966, 23 February. Second Ba'athist regime, more radical and leftist than its predecessor, seizes power in Syria following coup.

1967. Britain announces that it will cut forces abroad sharply.

1967, 5-10 June. Third Arab-Israeli (Six-Day) War.

1968. Ba'ath Party seizes power in Iraq.

1968. Britain announces withdrawal of its military forces from east of Suez.

1970, 13 November. Hafez el-Assad gains power in Syria following a bloodless coup.

1971, 18 July. United Arab Emirates established.

1971, 14 August. Independence of Bahrain.

1971, 21 September. Independence of Qatar.

1973, 6-24 October. Fourth Arab-Israeli (October) War.

1973, October. OAPEC oil embargo.

1975. Iraqi attempt to pressure Kuwait to cede territory.

1975. Gulf states seek to form a collective security grouping.

1976. Iraq obtains technology from France and Italy that enables it to embark on a nuclear weapons program.

1976, May. Renewed Kurdish revolt in Iraq.

1978, 14-21 March. Israeli punitive raid into southern Lebanon (Litani River operation) to destroy PLO bases and concentrations.

1978-1980. Iraq leads the Arab world in rejecting the Camp David accords and opposing the Egyptian-Israeli *rapprochement*.

1979, 16 January. Shah flees Iran.

1979, 12 February. Ayatollah Khomeini establishes Islamic Republic in Iran.

1979, 26 March. Camp David Accords (Egyptian-Israeli peace treaty).

1979, April. Arab League suspends Egypt's membership following Camp David Accords.

1979, 16 July. Saddam Hussein succeeds Hasan Bakr as president of Iraq, commander in chief, and Ba'ath Party chief.

1979, 3 November. US Embassy in Teheran, Iran, occupied by Iranian mob; embassy personnel seized and held hostage until 20 January 1981.

1980, 1/2 May. Failure of US military hostage rescue mission (Operation EAGLE CLAW) in Iran.

1980, 9 September-1988, 20 August. Iran-Iraq War. Kuwait and other Arab states provide financial aid to Iraq.

1981. Conservative Arab states on the Persian Gulf form the Gulf Cooperation Council (GCC), an organization to promote economic cooperation and collective security. The six member states are: Saudi Arabia, Kuwait, Bahrain, Qatar, the United Arab Emirates, and Oman.

1981, 7 June. Israeli aircraft bomb and destroy Iraq's Osirak nuclear reactor.

1982, 6 June-3 September. Israeli invasion of Lebanon (operation PEACE FOR GALILEE) and siege of Moslem West Beirut.

1982, 13 June. Crown Prince Fahd succeeds King Khalid as Saudi monarch.

1985. Attempted assassination of Kuwait's Sheikh Jabir al-Ahmad Al Sabah by Iraqi radical al-Dawa group.

1987, 17 May. *USS Stark* attacked, presumably mistakenly, by an Iraqi aircraft and damaged by a surface-skimmer antiship missile; 37 US sailors killed. Iraq apologizes for the attack.

1988, 3 July. Shoot-down of Iranian commercial jetliner by US Navy warship *Vincennes* in the Persian Gulf; all 290 on board the aircraft are killed.

1988, August-1990, August. Iraqi-Kuwaiti relations strained by territorial disputes, including the question of sovereignty over Bubiyan Island.

1988, December. Pan Am flight 103 blown-up by terrorist bomb over Lockerbie, Scotland; 270 are killed (including 11 on the ground); Iran linked to the blast.

1990, 2 August. Iraqi invasion of Kuwait.

APPENDIX B

The Desert Environment (Terrain and Climate)

The effects of terrain and climate on combat are basic considerations in the planning of military operations. Thus, it is necessary to outline the main geographical and climatological factors which would affect military operations leading to the liberation of Kuwait from Iraqi control.

The terrain of Kuwait and the neighboring areas of southern and southwestern Iraq and northeastern Saudi Arabia can be described as generally flat or gently rolling, and rising to the southwest. The vegetation is sparse and low, and trees are unknown except at oases or other permanent fresh water sources. The area's climate is generally warm and dry, and not unlike that of the Sahara, or the American southwest at low altitudes. As in those other desert areas, there is as great a range of temperature between day and night as between the average temperatures of summer and winter. Outside of Kuwait City and the agricultural regions along the Tigris-Euphrates river systems, the population is sparse and there are few settlements.

Terrain

Most Americans who have watched recent Western news programs have obtained an impression of the Arabian countryside. Sand, usually hard-packed, is pervasive. There are extensive areas which are nearly flat, with few significant terrain features, other than dry stream beds or water-courses called *wadis*, and salt-flats called *sabkhas*. Wadis, like arroyos in the American

Southwest, carry water only during rare rainstorms, and are dangerous only should a cloudburst and flash flood occur while someone is camped in the wadi itself. Most wadis descend from the higher country in the southwest and run east or northeast, toward the Persian Gulf and the Tigris-Euphrates river system. One major wadi, the Wadi el Batin, lies in Iraq just beyond the western Kuwaiti border, and is bounded on both its northwest and southeast banks by intermittent escarpments.

Sabkhas are widespread, generally lower than the surrounding terrain, and often collect rainwater. The of most sabkhas is soft, and can significantly limit mobility, especially of wheeled vehicles. There are often deposits of gypsum and calcite on the as well, and the dust which can rise during windy periods or by the passage of vehicles.

Within Kuwait itself, the country rises from the coast to the southwest; the highest point in the country, some 75 kilometers (47 miles) west-southwest of Al Jahrah, is 299 meters (about 980 feet) above sea level. Neighboring areas of Saudi Arabia and Iraq are slightly higher, but few areas likely to see combat are over 400 meters (1,312 feet) high.

The general surface is usually either gravel or firm sand, both trafficable to most military vehicles. There are, however, scattered patches of soft sand and some low rocky outcroppings. A modest erg (area of sand dunes) is in the At Tawal region of southwestern Iraq along the Saudi border, lying some 175 kilometers (110 miles) south-southwest of An Nasiriyah. Sabkhas are especially common along the northern and western shores of the Khalij al-Kuwayt (Kuwait Bay). Salt marshes cover much of Bubiyan Island and the Faw or Fao Peninsula.

Most of the vegetation is low and consists largely of scrub and salt grass. During the annual rains in early spring the desert undergoes a brief and colorful flowering of desert plants, but this lasts for just a few weeks. Only a few areas of Kuwait support denser vegetation. Al-Jahrah Oasis, and a few small fertile areas in coastal southeastern Kuwait support some trees, often date palms, and permit very limited cultivation. The Kuwaiti government created a "green belt," including fairly extensive areas of trees, around the southern edges of Kuwait City.

The marshy areas and the intensively irrigated areas of Iraq tied to the Tigris-Euphrates river system support much more vegetation, although there are no major wooded areas. The combination of waterways, permanently damp ground, and denser vegetation make these areas unsuitable to major mechanized operations and movements. The high water table also sharply limits the depth of entrenchments, and field fortifications in these water-rich areas will perforce be less extensive than elsewhere. The Iraqis discovered this effect, to their cost, in fighting on the Faw Peninsula during the Gulf War of 1980-1988.

The coastal islands at the northwestern end of the Persian Gulf, notably Bubiyan Island (Jazirat Bubiyan) and smaller Warbah Island (Jazirat Warbah) to the north (near the Iraqi naval base of Umm Qasr), are a special terrain area. Both regions, like the adjacent Iraqi coast, are low and marshy. Bubiyan Island has sand beaches along its southeastern coast, but otherwise the shoreline is tidally inundated saltwater marsh and mud flats. There are few if any permanent installations on either island, although there is a police post on the southeast corner of Bubiyan Island, and a disused oil storage tank in the south, at the end of a bridge and road which connects Bubiyan with the Kuwaiti mainland. Bubiyan Island is low and damp, rising no more than five meters (16 feet) above sea level. Warbah Island is likewise an overgrown sandbar, and there is a police post near its eastern end.

Faylaka Island (Jazirat Faylaka) and the neighboring islet of Jazirat Miskan lie outside the entrance to Kuwait Bay. Miskan is only 600 meters long, and supports a mere handful of buildings with no harbor or airfield facilities. Faylaka, however, is much larger and supports a town (Az Zawr) with a small anchorage, a helicopter pad, and other features. Although no part of either island is more than eight meters (26 feet) high, no areas are marshy or flooded. Consequently, either Faylaka or Miskan could prove useful militarily.

The coastline of mainland Kuwait is dry and generally low, while most of Iraq's 60-odd kilometers of shoreline is marshy, and much of it is fringed with mud-flats exposed at low tide. Most of the shoreline of Kuwait Bay is also fringed with mud flats; only a narrow stretch of shore at the extreme western end (near Al-Jahrah), and the harbors in Kuwait City, are easily accessible from the sea. Kuwait's coastline facing the Persian Gulf proper, however, is much better suited for landing operations, and the Kuwait-Dhahran highway runs within five kilometers of the coast.

Transport Infrastructure

The transport infrastructure of roads and railways in Kuwait and southeastern Iraq has not developed along European or American patterns. This is due largely to the effects of demography on geography. W.where there are few or no inhabitants, there are usually no roads. This is particularly the case in Kuwait, where nine out of ten people live in Kuwait City or one of the few other large towns nearby.

Kuwait has three major road axes. The first runs south from Kuwait City toward Dhahran. Branch and feeder routes serve the extensive oil facilities in southern Kuwait. A secondary road leads south-southwest to the Saudi border, and then due south to Al-Wari'ah, where it joins with the Saudi road-pipeline route from Dhahran northwest along the Iraqi border to Jordan. Kuwait's second major road runs west from Kuwait City through the Al Jahrah Oasis at the west end of Kuwait Bay (Jun al-Kuwayt). From there the road bends southwest across the Saudi frontier to Ar Ruq'i, and then turns south-southwest to meet the Dhahran-Jordan highway near Hafar al Batin, some 240 kilometers (150 miles) southwest of Kuwait City. The third primary road runs north from Al Jahrah to Basra in Iraq, tying in with the main Iraqi road and rail nets, as well as with the Tigris-Euphrates waterway system.

This last route is probably the most important, for it is the major overland route by which Iraqi forces in Kuwait are supplied and supported. In the event of hostilities, it will be necessary to interdict or sever that line of supply and communications.

Transportation in southeastern Iraq is complicated by the terrain, namely the Tigris-Euphrates waterway system. The area inside the triangle Basra-An Nasiriyah-Al 'Amarah is waterlogged, especially in spring. Rice is the principal crop, and the inhabitants (known as "marsh Arabs") are likelier to travel by boat or raft than on land. Land transport routes across that area are virtually non-existent. The main road runs from Faw through Basra, and then along the Tigris through Al 'Amarah and Al Kut to the Baghdad region. A secondary road follows the south bank of the Euphrates to An Nasiriyah and then leads on to Baghdad. A rail line follows a similar route, but there is few land connections between the two main transport corridors until they pass north and east of Al Kut and Al Duwaniyah, respectively. A major highway connects Al Kut and An Nasiriyah, and a secondary road connects An Nasiriyah with Al Qurnah, where the Tigris and Euphrates join to form the Shatt al'Arab. Basra is as much a key to transport routes in southeastern Iraq as Kuwait City is to the routes in Kuwait.

In addition to the Dhahran-Kuwait City-Basra highway, there is one other significant overland route between Saudi Arabia and Iraq. A secondary road leads south-southwest from An Najaf (145 kilometers, or 91 miles, south of Baghdad), crosses the Saudi border and intersects the Dhahran-Jordan highway at Rafha, and then travels south to Ha'il. Another secondary road leads south-southwest from As Samawah to As Salman, then west to Ash Shabakah, intersecting with the An Najaf-Rafha road. There also several tracks in the Saudi-Iraqi border area, but none of these are suitable for large scale military movements; they would have to be widened and surfaced to serve as logistical routes.

Climate

In its broad outlines, the climate of Kuwait and the surrounding country is similar to that of southern Arizona. Daytime temperatures are high, ranging from highs over 110 degrees F. in summer to perhaps 70 to 75 degrees in mid-winter. Nighttime temperatures are much lower, falling to the 70s in summer, and dropping near freezing in winter. Frosts in inland areas are comparatively frequent in winter, and standing water may freeze over on nights. Temperatures along the coast are moderated significantly by the warm waters of the Persian Gulf, and frosts are very rare in extreme southeastern Iraq and coastal Kuwait.

Rain amounts to between two and 18 centimeters (one to seven inches) annually, with a mean rainfall of 12 centimeters (just under five inches) at Kuwait City. Nearly all precipitation falls between November and April, mostly during the first three weeks of March, and this sudden and brief wet season causes the desert to bloom from mid-March to late April. Conditions in central and northern Iraq vary slightly from those in Kuwait, with the rainfall spread a little more evenly over the winter months. The Baghdad region receives a winter snowfall about one year out of three, and the highlands north and east of Mosul are generally snow-covered in January and February.

The prevailing wind, known as the *shamal*, generally blows from the west-northwest or northwest. In the winter these winds are cool, but from May to late September, they are hot and dry. During that season there can be gale-force shamals which may last for as long as a week, producing fierce sandstorms called *tauz*. In the spring and early summer there are also hot, dry winds from the south and southwest called the *suhaili*. A late summer wind known as the *qaus* sometimes blows from the southeast from July through October, bringing warm, damp conditions. Some winter winds blow from the northeast, bringing sharper cold, but these are more noticeable in central and northern Iraq than nearer the Gulf.

There are two notable windy periods during the year. The first two months of the year are affected by cold northerly winds from Siberia and Central Asia. These winds, which grow stronger as winter advances, create a great deal of dust and wind-blown sand, and produce a significant number of local sandstorms. During these storms, and in January and February generally, visibility is limited at or near ground level, and persistent conditions of airborne dust and sand have a significant effect on visibility, health and the effectiveness and maintainability of equipment.

A second "windy season" follows hard on the heels of the brief rains in March. From the third week of March (generally about the time of the Vernal Equinox) through the middle of April, the weather is often dominated by southerly suhaili (similar to the *sirocco* of North Africa or the *khamsin* of Egypt). Sometimes this produces such fierce dust- and sandstorms that the sky appears dusky even at noon. This wind and its associated effects would likely produce some generally non-serious casualties, but the effect on operational readiness for both personnel and equipment would be significant.

Men and Machinery in the Desert

Fortunately, there is substantial historical experience in desert campaigning to permit projections of the effects of the desert environment on men and machinery. Although the Palestinian and Mesopotamian campaigns of World War I contain some useful lessons, the main characteristics of modern mechanized warfare in the desert became apparent during operations in Libya, Egypt, and Tunisia during World War II. German General von Ravenstein, who commanded the 21st Panzer Division in Rommel's Afrika Korps in Libya and Egypt, aptly described the desert as "a tactician's paradise and a quartermaster's hell."

The lack of inhabitants and their accompanying infrastructure demands that armies operating in the desert build whatever facilities they need; there is virtually no possibility of appropriating existing civilian structures. During World War II, for example, the British extended the Egyptian railroad from Alexandria through Mersa Matruh and on to the Libyan-Egyptian border.

Since the desert is virtually barren of useful materials, everything an army needs must first be transported to the theater of operations, and then must be distributed by truck or aircraft to the units and their soldiers. With the proximity of the Saudi and Gulf state oil fields the fuel situation will be less critical than in the Western Desert during World War II. All the other sinews of war will have to be imported. The rates of material consumption of a modern army are immense, especially since nearly all the facilities needed will have to be built with imported materials.

As important as the logistical requirements imposed by the desert is the effect the environment on machinery. The dry, windy climate, coupled with sandy terrain, ensures that dust and grit get into virtually all machinery. This is especially damaging where surfaces lubricated with oil are exposed to air; the oil collects grit at an amazing rate, and the resulting abrasive paste dramatically increases rates of wear. For their operations in the Western Desert in 1941-1942, German records showed that the average tank engine could go only 3,500 kilometers between major overhauls, half the distance prevailing in continental Europe. Some motor vehicle engines wore out even faster, requiring overhaul and replacement four times as often as in normal European conditions. Likewise, gun barrels and tank tracks suffered increased wear. In all cases, soldiers had to pay close attention to the state of their equipment, and conduct constant maintenance to ensure the availability of items when needed.

In general, tracked vehicles are better suited for desert operations than wheeled vehicles. However, track systems require greater maintenance in the desert than elsewhere. The Israeli Defense Forces, which have accumulated a great deal of desert operational experience, prefer to employ their U.S.-built tanks (with live-tensioned track suspension systems) in the relatively sandy soils of the Sinai. The Israelis have generally committed their British-built tanks (with dead-link track) on the boulder- and rock-strewn Golan Heights area. The new Israeli-built Merkava tank has a track suspension system including the best features of live-tensioned and dead-link tracks, and so is more versatile.

In the Western Desert in World War II, the effect of the environment on aircraft operations and maintenance was as significant as that on vehicles. It is logical to assume that, with most modern combat aircraft powered by jet engines, such effects will be at least as great as in 1940-1943. The widespread use of helicopters by modern armies will undoubtedly also bring problems when such units operate in the desert. One notable historical example from recent history was the abortive U.S. attempt to rescue the hostages from Teheran in 1980 during the Iranian Revolution. The major problem, prior to the fiery crash at the "Desert One" forward airfield, was that the conditions of heat and dryness over the Iranian desert reduced the lift and flight capacities, and produced higher fuel consumption rates, among the Marine Corps helicopters. Similar conditions in Arabia will mean that helicopters will have reduced loiter time over the battlefield, reduced flight endurance, and reduced range. All of these factors will exert a considerable influence not only on attack helicopters and airmobile operations, but also on supply and medical evacuation. Operations in the cooler months would reduce the impact of the heat factor considerably, but dust-related wear would still be a problem.

The dryness of desert environments, and the related increase in overall static electricity, is a problem for electronic equipment. Sand and dust can also interfere with circuits and contribute to degradation of insulation. Static electricity could have serious consequences for such practices as "hot" refueling of aircraft and helicopters, and arrangements for secure grounding will have to be especially thorough to prevent mishaps.

Most of this wear and tear is due to virtually omnipresent clouds of dust. Dust clouds arise around any mechanical activity, including vehicle movement of any significant scale. Artillery fire also raises clouds of dust around the firing weapon (from recoil and blast), as well as in the impact area. Because of these conditions, battlefields often became obscured by thick clouds of dust and airborne sand. This limited visibility hampers communications and control, reduces unit cohesion, and increases the mechanical difficulties even under non-battle conditions.

The rigors of the desert apply to men no less than to their machines. The extremes of temperature and the lack of water make life difficult, and the dust and sand can make even the hardiest soldier miserable. Both Allied and Axis forces usually managed to maintain adequate water supplies in World War II, that effort required careful planning. Water supply can be a significant problem, and even under the best circumstances requires considerable effort to meet water supply needs. Troops in desert environments require an average of twice as much water as soldiers operating in temperate climates. Local water sources are generally inadequate, and supplies must be imported.

Many soldiers in World War II found the emptiness of the desert and its lack of landmarks disorienting, and there were numerous cases of units getting lost. Even reliance on dead reckoning and celestial navigation was not always sufficient, although specially-trained desert reconnaissance units—specifically the British Long-Range Desert Group—did remarkably well, in part because many of their personnel had prewar desert experience. The extremes of climate require special clothing arrangements.

The problem of land navigation is caused by the lack of familiar and identifiable landmarks; it is often difficult for a non-native to note the horizon line, which adds to the disorientation already noted. Further, during those periods when the dust subsides, particularly at night, the very clarity of desert air, free from haze and clouds, contributes to difficulty in estimating distances, and distant objects seem much closer than they really are. To add further to the miseries of the desert, poisonous spiders, scorpions, and snakes are common, and other pests, like fleas, ticks, and lice, abound as well.

Environment (Terrain and Climate)

	Jan	Feb	Mar	Apr	May	Jun	Jul	Aug	Sep	Oct	Nov	Dec
Prevailing Wind direction:												
	NE	NE	NE	NW	NW	NW	NW	NW	NW	NW	NW	NW
	NW	NW	NW	S	S	S	SW	SW	SW	SW		
Temps (Fahrenheit):												
High	75	77	88	100	107	115	112	117	107	95	85	78
Mean	55	56	64	73	83	93	91	94	82	72	62	58
Low	36	36	40	47	60	71	70	71	57	48	40	37
Rain (cm)	2	2	6	0.5	-	-	-	-	-	-	0.5	1

FIGURE 22. CLIMATE TIME-LINE (KUWAIT AND SOUTHEASTERN IRAQ)

These figures should be taken as general indicators only. Yearly and daily specific temperatures will vary considerably, and the highs in summer and lows in winter should be taken to indicate relatively extreme temperatures.

Conclusion

In light of these climatic conditions, the winter months from mid-October to late February are the most favorable for military operations. Because of the northerly winter wind, it would be best to undertake operations in the mid- to late autumn (October through December.) The summer months, or the period from late March through late August, would be least favorable for operations because of extremely high daytime temperatures. The first three weeks of March are often a period of good weather and exceptionally clear visibility, although the rains are sometimes heavy and can cause flooding of wadis.

Any operations undertaken in the region will have to make adequate logistical preparation for the water requirements of personnel. Many personnel may suffer mild respiratory irritation from the dryness of the atmosphere, exacerbated by airborne dust.

Operations in virtually any season will have to take account of visibility problems caused by airborne dust and sand. This is particularly severe during the warmer months, when heating of the desert floor by sunshine causes low-level air currents which suspend airborne particles near the ground and greatly limit visibility. This situation disappears during evening and nighttime, and is a major problem only from mid-morning until an hour or so before sunset, when lowered temperatures reduce the ground-level convection. This "daily dust problem" is in addition to any wind-generated sandstorms which may occur, and merely adds to the dust problems associated with mechanical activity.

APPENDIX C

Equipment and Weapons of the Forces*

I. United States

Infantry Weapons Score**

M16A1 Assault Rifle	0.32
M60 Machinegun	0.40
Squad Automatic Weapon (SAW)	0.77
M2HB .50 Cal MG	1.03
M203 GL	3.90
60mm Mortar	21.00
81mm Mortar	43.00
81mm Mortar (SP, mounted on M113 or LAV)	45.00
4.2" Mortar (SP, mounted on M113)	79.00
M113 APC	2.69
LVTP7	4.50
MK19 AGL	58.00

Antiarmor Weapons

LAW	5.00
M136 (AT4)	31.00
Dragon	34.00
M901 ITV	205.00
TOW HUMV	129.00
TOW Manpack	108.00

Artillery

M198 — 155mm (T)	223.00
M109A2 — 155mm (SP)	223.00
M110A2 — 203mm (SP)	173.00
MLRS	311.00

Air Defense Weapons

Stinger	16.00

---

* Principal sources: Jane's Weapons Systems series; IISS, *Military Balance*; Isby, *Armies of NATO's Central Front*.
** Scoring is based upon T.N. Dupuy's Operational Lethality Index (OLI) method; see *Numbers, Predictions, and War*, New York, 1979.

Chaparral	135.00	M1A1	1,049.00
Vulcan (T)	179.00		
Vulcan (SP)	187.00	**Aircraft***	
Improved Hawk	100.00	AH-1J Cobra	96.00
Patriot	122.00	AH-64 Apache	256.00
		AV8B Harrier	351.00
Armor		A-7	707.00
LAV (25mm)	96.00	F-16	1,359.00
M728 Engineer Vehicle	315.00	A-6E	1,484.00
Sheridan	506.00	A-10	2,697.00
M2 Bradley	534.00	F-14A	N/A
M3 Bradley		F/A-18	N/A
(Cavalry Version)	597.00	F-111F	1,922.00
M60A3	650.00	B-52	10,063.00
M1	984.00	F-117	N/A

II. United Kingdom

Infantry Weapons Score		MLRS	311.00
L1A1 Rifle	0.06		
L85 MG	0.24	**Air Defense**	
L86 MG	0.50	Blowpipe	17.00
GPMG	0.64	Javelin	20.00
M2HB .50 Cal MG	1.03	Rapier (SP)	133.00
51mm Mortar	11.00		
81mm Mortar	43.00	**Armor**	
81mm Mortar (SP)	45.00	Ferret	0.61
FV432	0.78	Scimitar	51.00
		MCV-80	52.00
Antiarmor		Scorpion	135.00
LAW-80	9.90	Challenger	925.00
Carl Gustav	31.00		
MILAN	82.00	**Air**	
Striker Swingfire	204.00	Lynx/TOW	60.00
		Harrier	351.00
Artillery		Jaguar1	536.00
M109A2 — 155mm (SP)	223.00	Tornado	1165.00

---

* Aircraft scores are for ground attack roles. Aircraft without scores are those in an air superiority role only.

III. France

Infantry Weapons Score
FA MAS Rifle	0.38
AA 53 MG	0.58
M2HB .50 Cal MG	1.03
81mm Mortar	43.00
120mm Mortar	100.00
VAB 6x6 (12.7mm)	4.60

Antiarmor
LRAC 89	20.00
MILAN	82.00
VABHOT	217.00

Artillery
TR — 155mm (Towed)	219.00

Air Defense
RH202 20mm	63.00
Mistral	30.00
Crotale	176.00
AMX-30 DCA	185.00

Armor
AMX-10RC	560.00
AMX30B2	766.00

Air
Gazelle HOT	46.00
Mirage F1	343.00
Mirage 2000	412.00
Jaguar	536.00

IV. Kuwait

(Though much of the Kuwaiti Army was destoyed, some men and equipment were able to escape into Saudi Arabia. The Kuwaiti government in exile is presently engaged in a large scale arms acquisition program.)

Infantry
FN-FAL	0.06
FN-MAG	0.64
M2HB .50 Cal MG	1.03
M113	2.69
Saracen	0.74
Fahd APC	3.83
81mm Mortar	43.00
120mm Mortar	85.00

Antiarmor
106mm RR	67.00
Vigilant	81.00
TOW	108.00
HOT (SP)	217.00

Artillery
M101 — 105mm	122.00
F3 — 155mm	243.00
M109A2 — 155mm (SP)	223.00
FROG-7	227.00

Air Defense
SA-7	16.00

Armor
Ferret	0.61
Saladin	102.00
BMP-2	414.00
Vickers Mk1	500.00
Chieftain	782.00
M84 (Yugoslavian T-72 on order)	977.00

Air
Gazelle (20mm)	9.04

A-4KU	342.00	Mirage F1	343.00

V. Saudi Arabia

Infantry Weapons		GCT	243.00
G3 Assault Rifle	0.24	Astros II	193.00
FN-MAG	0.64		
MG3 MG	0.76	**Air Defense**	
M2HB .50 Cal MG	1.03	Redeye	13.00
M113	2.69	Stinger	16.00
81mm Mortar	43.00	M42	62.00
107mm Mortar	75.00	M117	117.00
81mm Mortar (SP)	45.00		
107mm Mortar (SP)	79.00	**Armor**	
		M3 Panhard	3.31
Antiarmor		BMR600	4.07
75mm RR	25.00	AML60	31.00
90mm RR	34.00	UR416	51.00
AMX10-HOT	235.00	AMX-10P	126.00
Dragon	34.00	M60A1	622.00
TOW APC	205.00	M60A3	643.00
		V-150/20	102.00
Artillery		AMX30S	868.00
M56 — 105mm	113.00		
M101 — 105mm	103.00	**Air**	
M102 — 105mm	105.00	F-16A	717.00
M198 — 155mm (T)	223.00	F-5E	134.00
M109A2	223.00	Tornado	1165.00
FH-70	223.00	F-15C	N/A

VI. Egypt

Infantry		120mm Mortar	85.00
AK47/AKM	0.19	160mm Mortar	102.00
RPD	0.21	M113A2	2.69
SGM MG	0.44	BTR50	0.75
M60 MG	0.44	BTR60P	1.78
FN-MAG	0.64	BTR60PB	3.03
DShK	1.04	OT62	3.07
M2HB .50 Cal MG	1.03	Fahd	3.83
82mm Mortar	35.00	Walid	

Antiarmor

RPG-7V	18.00
B-11 107mm RR	59.00
AT-3 (Sagger)	88.00
AT-3 on Jeep	97.00
MILAN	82.00
106mm RR on Jeep	70.00
Swingfire	100.00
TOW	108.00
BRDM W/Sagger	163.00
M901 ITV	205.00

Artillery

D74 — 122mm How	198.00
D30 — 122mm How	190.00
M1938 — 122mm	190.00
M109A2 — 155mm	223.00
BM-21 — 122mm MRL	404.00
VAP-80 — 80mm MRL	179.00
BM-24 240mm MRL	179.00
M-51 — 130mm MRL	298.00
BM-13-16 — 132mm MRL	273.00
BM-14-16 — 140mm MRL	251.00
FROG-7	227.00

Air Defense

SA-7	16.00
ZPU2	1.75
ZPU-4	2.42
ZSU-57-2	171.00
ZSU-23-4	177.00
S-60	109.00
SA	999.00
SA2	82.00

Armor

BMR-600 IVF	4.07
BMP1	286.00
SU100	199.00
T62	691.00
M60A3	643.00

Air

Gazelle (HOT)	46.00
Gazelle (20mm)	9.00
Mirage 5E2	614.00
F-4E	720.00
J-6	300.00
Alphajet	297.00
Mig-17	144.00
Mig-21	N/A
J-7	N/A
F-16A	N/A
F-16C	N/A
Mirage 5E	N/A
Mirage 2000C	N/A

VII. Syria

Infantry

AK47/AKM	0.19
RPD MG	0.21
SGM MG	0.44
DShK MG	1.04
BTR50PU	0.75
OT64C	2.63
BTR60PB	3.03
82mm Mortar	35.00
120mm Mortar	85.00
160mm Mortar	102.00
240mm Mortar	72.00

Antiarmor

RPG-7	18.00
SPG-9	28.00
AT-3 Sagger ATGM	88.00
BRDM AT-3	163.00
AT-3	88.00
MILAN	82.00

AT-4	87.00
Artillery	
122mm SP	192.00
152mm SP	216.00
D-30 122mm	190.00
M-46 130mm	248.00
D-1 152mm	173.00
BM-21 MRL	
— 122mm MRL	404.00
BM-27 — 220mm MRL	299.00
BM-24 — 240mm MRL	179.00
FROG-7	270.00
Air Defense	
S-60	109.00
KS-19	144.00
ZSU-57-2	171.00
ZSU-23-7	177.00
SA-7	16.00
SA9	99.00
SA13	147.00
Armor	
BRDM-2	3.61
BMP-1	286.00
T-62A	691.00
Air	
Mi-24D / Mi-25	108.00
Gazelle (HOT)	46.00
MiG-17	144.00
Su-7	82.00
Su-20	292.00
MiG23BN	350.00
MiG-21	N/A
MiG-25	N/A
MiG-23	N/A
MiG-29	N/A

VIII. Iraq

Infantry Weapons	
AK47/AKM	0.19
RPD MG	0.21
SGM MG	0.44
DShK MG	1.04
81mm Mortar	43.00
120mm Mortar	85.00
120mm Mortar (SP)	93.00
160mm Mortar	102.00
160mm Mortar (SP)	112.00
BTR152	0.98
BTR50P/PK/PU	0.75
BTR60PB	3.03
OT62B	3.07
OT-64C(1)	4.20
OT-65A	0.84
EE-11 IFV	3.00
Antiarmor	
RPG-7	18.00
SPG-9	28.00
B-10	30.00
AT-1 BRDM	76.00
AT-2 BRDM	143.00
AT-3	88.00
AT-3 BRDM	159.00
AT-4	87.00
AT-5 BRMD	268.00
VCR HOT	217.00
MILAN	82.00
Artillery	
D-30 — 122mm	190.00
M-46 — 130mm	248.00
D1 — 152mm	173.00
D-20 — 152mm	199.00
M115 — 203mm	151.00
FH-70 — 155mm	223.00
G5 — 155mm	233.00
M109A1 — 155mm SP	233.00

Equipment and Weapons of the Forces

GCT — 155mm	243.00	SA-6	154.00
Type 59-1 — 130mm	210.00	Roland	118.00
GH N-45 — 155mm	233.00		
56 Pack How — 105mm	113.00	**Armor**	
Type 63 — 107mm	172.00	BRDM-2	3.61
BM-13-16 — 132mm	273.00	BMP	1286.00
2S1 — 122mm SP	192.00	BMP	2414.00
2S3 — 152mm SP	216.00	PT-76	94.00
M109 — 155mm SP	223.00	AML60	31.00
GCT — 155mm	243.00	AML90	148.00
BM-21 MRL		T-54/55	431.00
— 122mm MRL	404.00	Khalid	787.00
Majnoon — 155mm SP	256.00	Type 59	365.00
Al Fao — 155mm	183.00	Type 69-II	423.00
FROG-7	270.00	T-62	691.00
Astros II MRL	193.00	T-72	977.00
		M3 Panhard Chieftain	782.00
Air Defense			
ZPU-1	1.17	**Air**	
ZPU-2	1.75	Mi-24 (Hind D)	108.00
ZPU-4	2.42	Gazelle	46.00
ZU-23	116.00	BO-105	70.00
KS-12	80.00	Su-25	495.00
KS-19	144.00	F1-C-200	343.00
KS-30	214.00	Su-7	82.00
ZSU-57-2	171.00	Su-20	292.00
ZSU-23-4	177.00	Su-25	495.00
S-60	109.00	J-6	300.00
ZU-23	116.00	MiG-21 SMB	N/A
GDF-001	234.00	MiG-25 (Foxbat-E)	N/A
M39	47.00	MiG-29	N/A
M53-59	141.00	J-7	N/A
SA-2	82.00	Mirage F-1EQ	N/A
SA-3	71.00	Su-24 (Fencer-D)	424.00
SA-13	147.00	Tu-16 (Badger)	2259.00
SA-9	99.00	Tu-22 (Blinder-A)	1122.00
SA-8A	96.00		

Organization of Opposing Forces

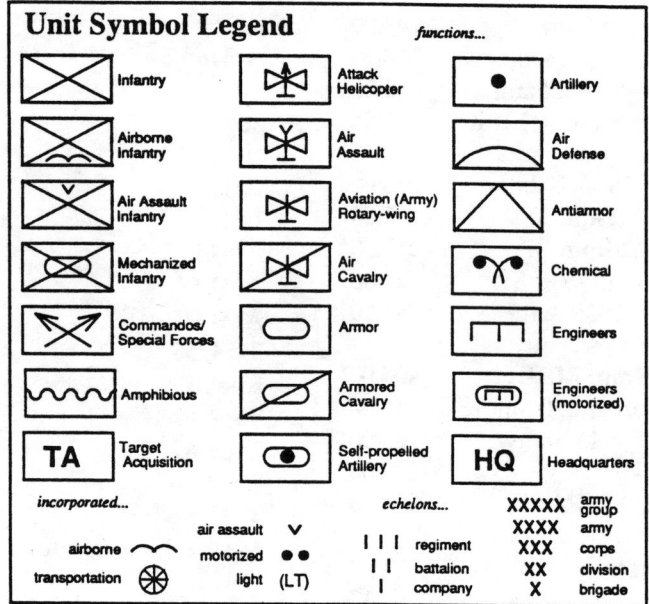

FIGURE 23. UNIT SYMBOL LEGEND

US Mechanized Division

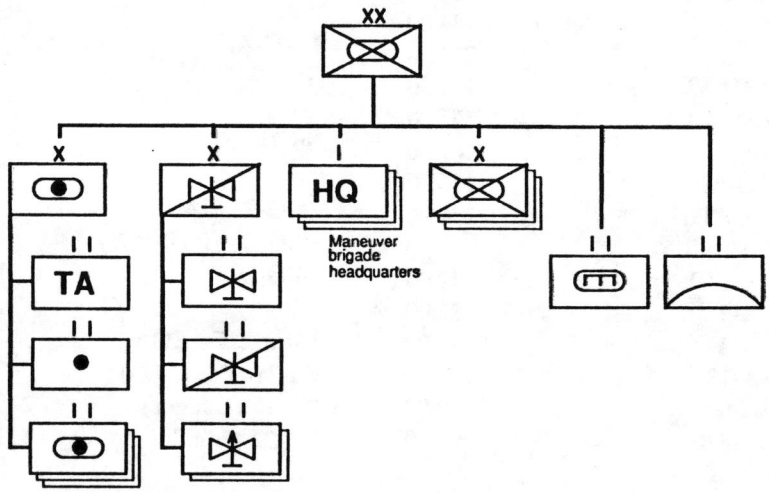

FIGURE 24. US MECHANIZED DIVISION

APPENDIX D

Organization of Opposing Forces*

I. United Nations Forces

A. United States Forces

Army
- XVIII Airborne Corps
  - 82d Airborne Division
  - 24th Mechanized Infantry Division (-)
    - 197th Mechanized Infantry Brigade
  - 101st Air Assault Division
  - 1st Cavalry Division (-)
    - 2d Brigade, 2d Armored Division
  - 3d Armored Cavalry Regt
  - XVIII Corps Artillery
  - 18th Aviation Brigade
  - 11th Air Defense Artillery (ADA) Brigade
  - Corps troops
- VII Corps
  - 1st Armored Division

---

* The following sources—not always consistent—have been consulted: *Jane's Defence Weekly*; IISS, *The Military Balance*; "For Your Eyes Only"; French Embassy, *France Update*; Washington *Post*; Washington *Times*.

US Armored Division

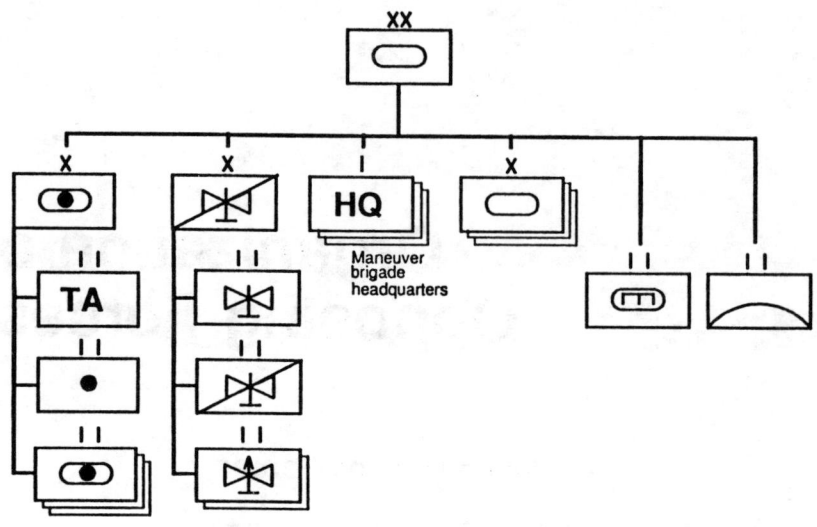

FIGURE 25. US ARMORED DIVISION

US Air Assault Division

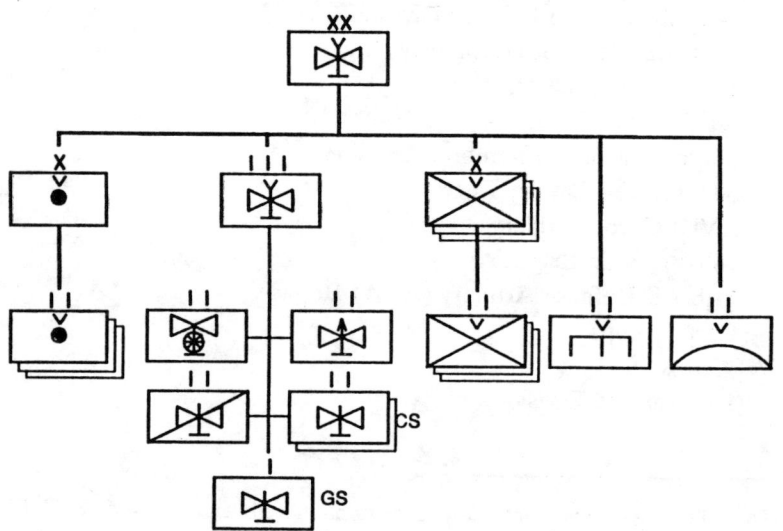

FIGURE 26. US AIR ASSAULT DIVISION

3d Armored Division
1st Mechanized Infantry Division
2d Armored Division**
VII Corps Artillery
12th Aviation Brigade
Corps Troops
c. 250,000 troops
c. 2,000 tanks
c. 900 artillery guns
c. 400 combat helicopters

Marines

1st MEF (Marine Expeditionary Force)***
    1st Marine Division
    2d Marine Division
4th MEB (Marine Expeditionary Brigade)(Afloat)
5th MEB (Afloat)
c. 70,000 ground combat troops (plus 40,000 in 2 air wings—2d, 3d—and other support)
c. 250 tanks
c. 300 artillery guns
c. 100 combat helicopters
c. 450 ground attack aircraft

Navy

  2 Battleships
  6 Carriers (3 forward, 3 back)
40 Other warships
c. 140 Combat aircraft (another 180 combat a/c in reserve, with 3 uncommitted carriers)

** This headquarters appears to have been used to establish a provisional mechanized infantry division.

*** An MEF usually consists of a Marine division, a Marine air wing, and other elements. In this case, one MEF contains two divisions and two air wings. The MEBs shown are not components of the MEFs.

US Airborne Division

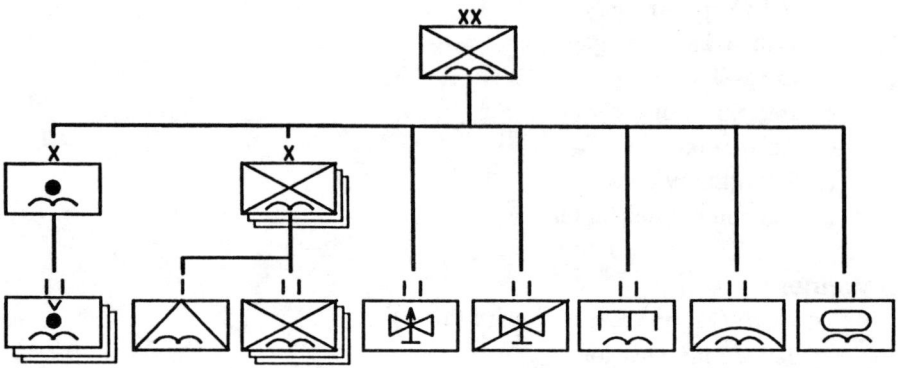

FIGURE 27. US AIRBORNE DIVISION

U.S. Marine Corps Division

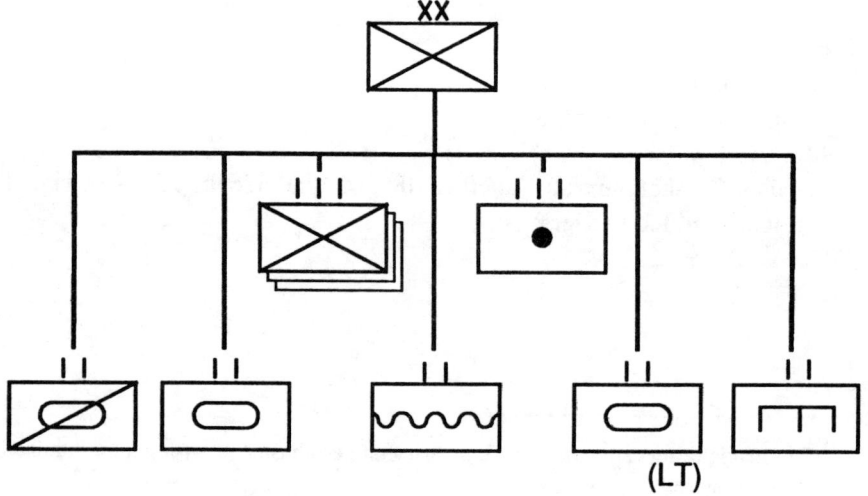

FIGURE 28. US MARINE CORPS DIVISION

Air Force
- c. 800 Combat aircraft
- 6 Air superiority squadrons
- 30 Ground attack squadrons
- 20 Support squadrons
- 3 Strategic bomber squadrons

B. United Nations Allies

Army
United Kingdom
    1st Armored Division
France
    6th Light Armored Division
Egypt
    3d Armored Division
    3d Mechanized Infantry Division
Syria
    9th Armored Division
Saudi Arabia
    I Corps
        4th Armored Brigade
        8th Armored Brigade
        20th Mechanized Infantry Brigade
        U/I Brigade (unidentified)
        National Guard Brigade
Kuwait
    35th Armored Brigade (Under Saudi I Corps)
        (Another armored brigade has been reformed, but with a limited number of tanks. Approximately 300 M84 (T-72 derivative) tanks are reported to be on order from Yugoslavia.)
Contingents from: Bahrain, Bangladesh, Morocco, Oman, Niger, Pakistan, Qatar, Senegal, the United Arab Emirates.
- c. 150,000 Personnel
- c. 8 divisions/division equivalents
- c. 1,200 tanks
- c. 800 artillery guns
- c. 150 combat helicopters

British 1st Armored Division

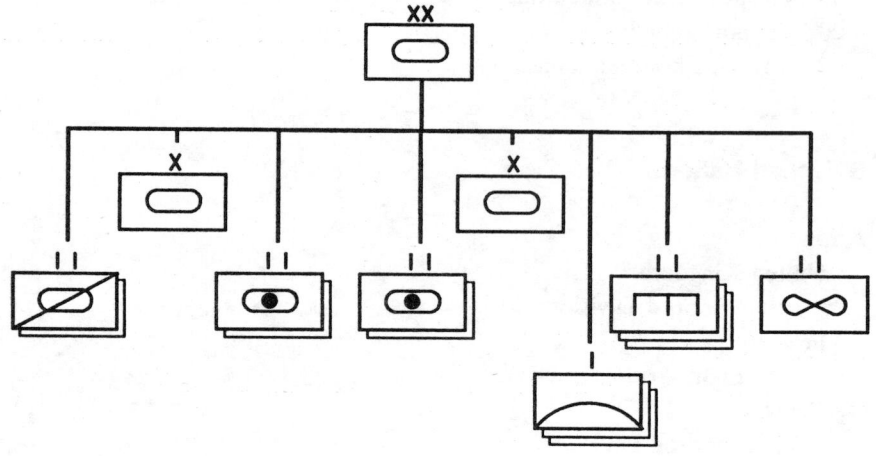

Figure 29. British 1st Armored Division

French 6th Light Armored Division

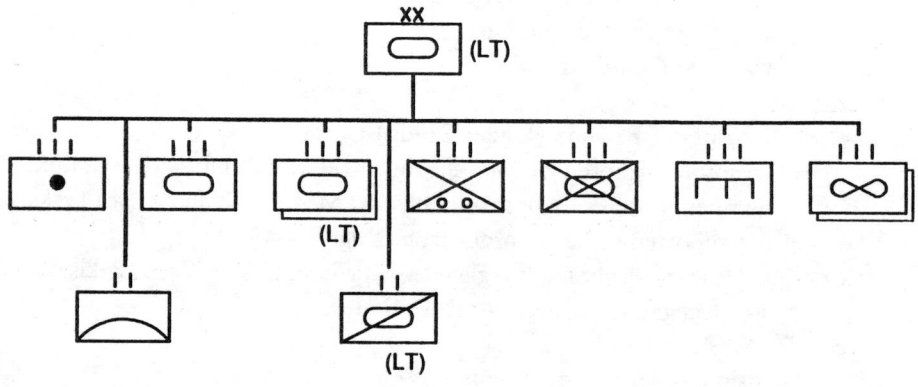

Figure 30. French 6th Light Armored Division

Air Forces
c. 500 Combat aircraft

Contingents from: Bahrain, Canada, Egypt, France (c. 200), Italy, Kuwait, Oman, Qatar, Saudi Arabia, United Arab Emirates, United Kingdom (c. 100). In addition, there are NATO contingents from Germany, Belgium, and Italy, based in Turkey.

Navy

Warships:	36
Mine-hunters:	5
Landing ships:	2
Support ships:	14

Contingents from: Argentina, Australia, Belgium, Canada, Denmark, Norway, France (7 warships, 1 other), Greece, Italy, the Netherlands, Portugal, Spain, the United Kingdom (7 warships, 11 others), the Soviet Union.

II. Iraqi Forces

A. Overall

Army

Personnel: c. 1,000,000

- 7 Corps
- 3 Armored divisions (Republican Guard)
- 8 Mechanized infantry divisions (3 Republican Guard)
- 42 Infantry divisions
- 21+ Special Forces (SF) brigades (1 Republican Guard)
- 2 Surface-to-Surface (SSM) brigades

Tanks:	5,500 (1,000 T-72; 1,500 T-62, 3,000 T-54/55/59/60)
APC/IFV:	7,500
Artillery (Towed):	3,000
(Self-propelled):	500
(Multiple Rocket Launchers):	200
Attack Helicopters:	160

Egyptian Mechanized Division

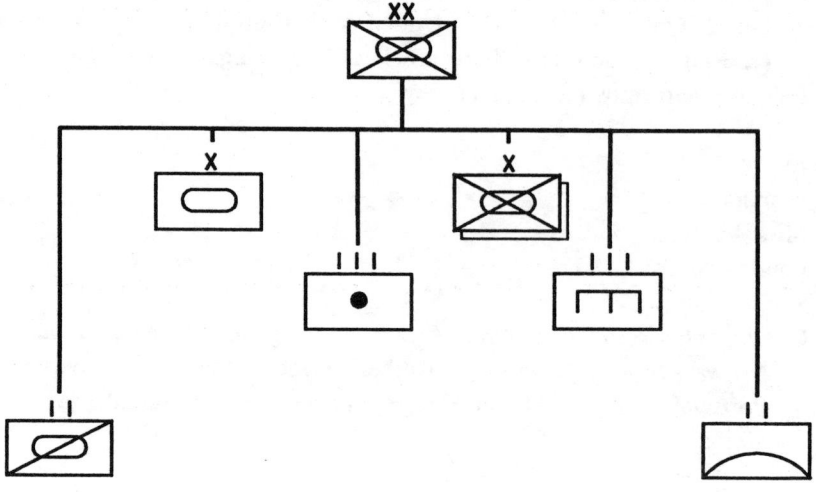

FIGURE 31. EGYPTIAN ARMORED DIVISION

Egyptian Armored Division

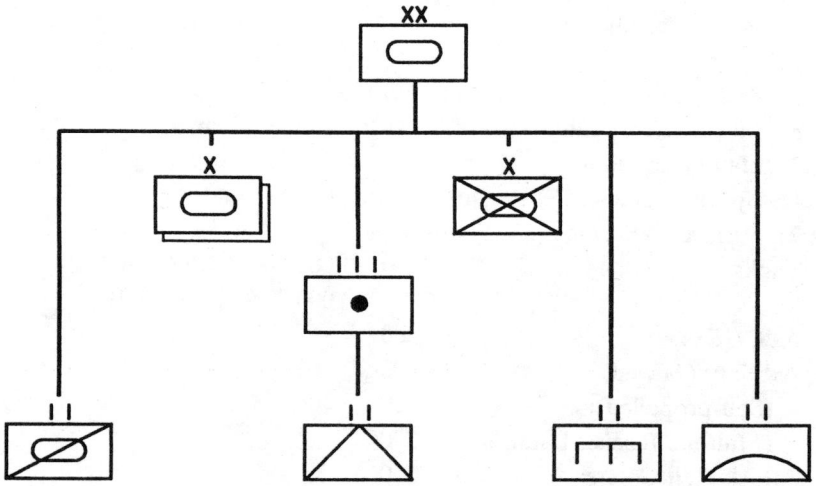

FIGURE 32. EGYPTIAN MECHANIZED INFANTRY DIVISION

Air Defense: guns:	4,000
Surface to air missiles (SAM):	
Long Range (Scud-B)	36
Medium Range (Frog)	30
Other	600

Air Force
Personnel:	40,000
Combat aircraft:	675
Bombers:	2 squadrons
Fighter/Ground Attack:	22 squadrons
Fighter:	17 squadrons

Navy
Personnel:	5,000
Base: Umm Qasr	
Frigates:	5
Corvettes:	6
Missile & Torpedo craft:	14
Inshore Patrol:	20
Amphibious:	6 (3 LST, 3 LSM)

Paramilitary
Personnel: c. 850,000 (People's Army)

B. Southern Front

Army
Personnel: 500,000
Corps: 5: I (Republican Guard), II, III, IV, VIII
Divisions: 28 (3 armored, 8 mechanized, 15 infantry (4 motorized), 1 marine, 1 special forces - in Kuwait City)

Tanks:	4,000	
APC/IFV:	4,000	
Artillery:	2,000	guns
	150	MRL
Air Defense:	2,000	guns
	300	SAMs

Syrian Armored Division

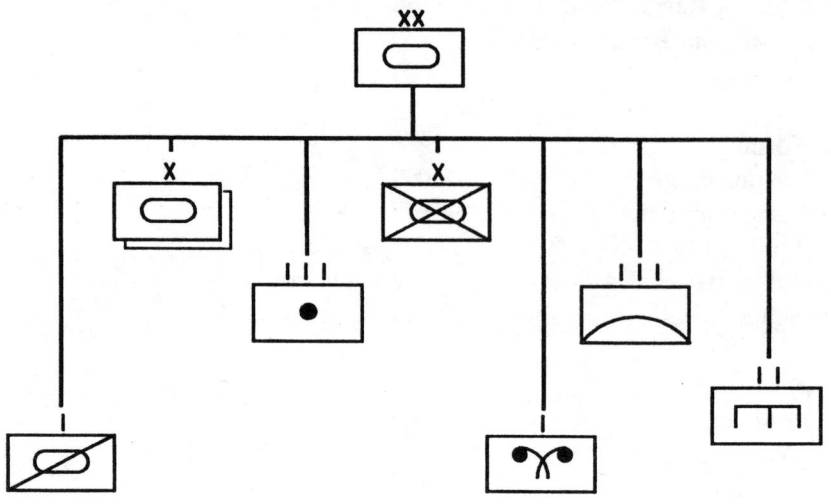

FIGURE 33. SYRIAN ARMORED DIVISION

Iraqi Armored Division

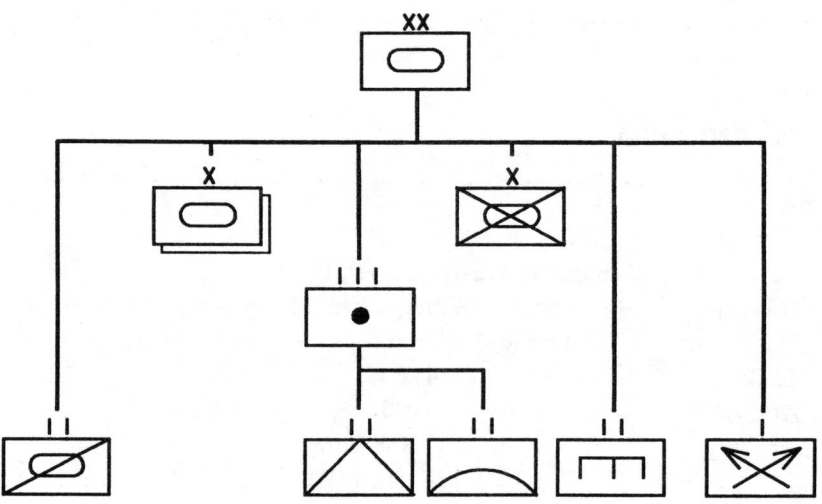

FIGURE 34. IRAQI ARMORED DIVISION

Deployment:
- a. In defensive positions along Saudi border west of Kuwait: 100,000
- b. In defensive positions in southern and central Kuwait: 150,000
- c. Along Kuwait coast: 130,000
- d. Republican Guards (I Corps, 3 armored divisions, 3 mech infantry divisions) between Basra and Kuwait border: 100,000

Air Force
Combat aircraft: 500
                   30 squadrons
Personnel: 30,000

Iraqi Mechanized Division

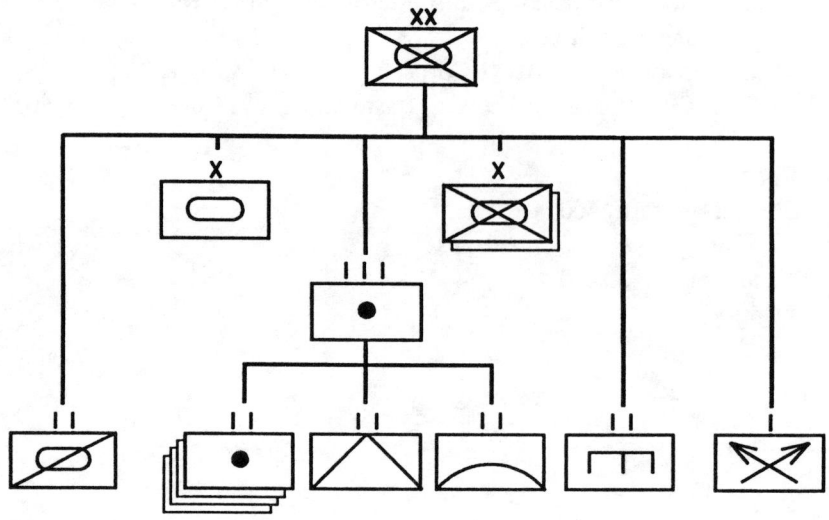

FIGURE 35. IRAQI MECHANIZED INFANTRY DIVISION

Iraqi Infantry Division

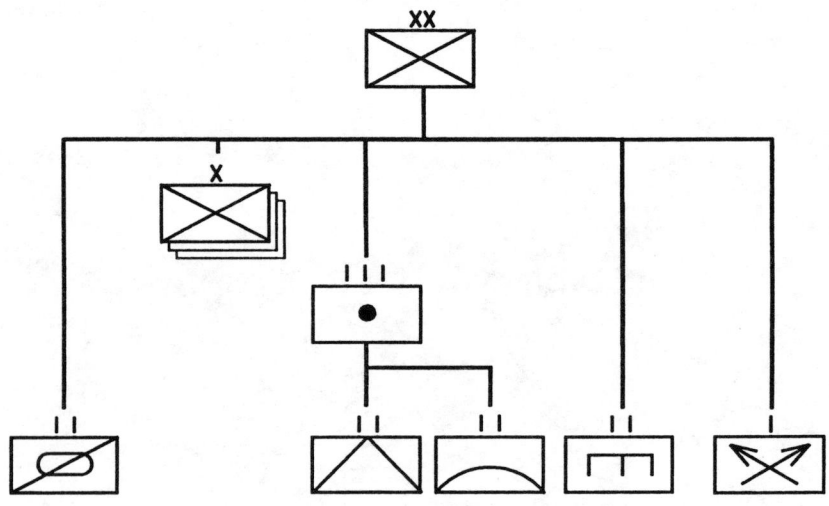

FIGURE 36. IRAQI INFANTRY DIVISION

APPENDIX E

Relative Combat Effectiveness

Anyone who has studied only a little military history is bound to note that some armies have been able consistently to outperform their opponents. This was true of the German Army in comparison with its opponents in both World War I and World War II. It has been true of the Israeli Army in comparison with Arab opponents in the several brief periods of intensive Arab-Israeli conflict from 1948 through 1982. But, while a difference in the quality of military performance has been clear, it is only recently that a methodology has been developed which would permit a reliable measurement of the undeniable, but not easily quantifiable, qualitative difference between these excellent armies and the less skillful opposing military forces.

By using the Tactical Numerical Deterministic Model (TNDM), it is possible to assess the relative combat effectiveness of opposing forces in historical battles and engagements consistently and scientifically. The TNDM process involves the calculation of two ratios.

The *Combat Power Ratio* is based on the strength, weapons and circumstances of opposing forces in a battle.

The *Result Ratio* is based upon the outcome of the battle.

The *Combat Power*, P, of each opponent is obtained by calculations in which factors representing the effects of the environmental and operational circumstances are applied to the Force Strengths of each side. The Combat Power Ratio is obtained by dividing the P for one side, P_1, by the P for the other side, P_2. The ratio P_1/P_2 defines the *theoretical* outcome of a battle. If P_1/P_2 is greater than 1.0, then Side 1 should be, or should have been, successful; if the ratio is less than 1.0, then Side 2 should theoretically win, or have won.

The actual outcome of a historical battle is based upon a comparison of the calculated outcome or Result for each side. This value (R) is calculated by adding an expert assessment of the degree of mission accomplishment (ranging from 1 to 10) with a calculation of spatial effectiveness (ability to gain or hold ground, usually ranging from about +4 to -4) and casualty effectiveness based on efficiency related to casualties incurred (also ranging from about +4 to -4). The outcome is then described by a Results Ratio, R_1/R_2. If R_1/R_2 is greater than 1.0, Side 1 actually won; if R_1/R_2 is less than 1.0, Side 2 was successful. Unless the two sides were perfectly matched in capability (something which occurs infrequently) P_1/P_2 is practically never identical to R_1/R_2, although it may be close.

The *Relative Combat Effectiveness Value*, CEV, of the opposing sides is defined as a comparison of actual performance with theoretical performance, or the ratio of R_1/R_2 to P_1/P_2, or:

$$CEV_1 = (R_1/R_2)/(P_1/P_2)$$

At the Battle of Austerlitz 1805, a TNDM analysis shows that the Combat Power Ratio of Napoleon's outnumbered French army with respect to the Austro-Russian allied army was: $P_{fr}/P_{al} = 0.94$. Before the battle an objective observer would have expected that Napoleon would have been defeated in a close-fought battle. However, the Result Ratio for Napoleon's overwhelming victory in this battle was $R_{fr}/R_{al} = 2.02$. The CEV for the French is calculated as follows:

$$CEV_{fr} = 2.02/0.94 = 2.15$$

In other words, the quality of Napoleon's leadership combined with the excellence of his Grand Army meant that 100 French troops were the equivalent of more than 200 comparably-equipped allied troops.

In World War I and World War II the German Army consistently outperformed British, French, and American opponents. But not by much. While the differential value varied from one battle to another, and from one set of opponents to another, the CEV ratio of ratios—$(R_g/R_a)/(P_g/P_a)$—averaged about 1.2. In other words, 100 Germans were roughly the equivalent of 120 comparably equipped American, British, or French soldiers. This was true when the Germans were attacking, when they were defending, when they won, and when they lost. On the Eastern Front, in both of these wars, the

CEV ratio was much higher in favor of the Germans, generally ranging between values of 2.0 to 3.0. Or, 100 German soldiers were the combat equivalent of 200 or more Russian soldiers.

Why didn't the Germans win? They were outnumbered in the West by more than 2.0 to 1.0. They were outnumbered by the Russians in World War II by more than 3.0 to 1.0.

In the Arab-Israeli Wars the Israelis have similarly outperformed their Arab foes. While this has varied from one Arab army to another (the Jordanians giving the best performance against the Israelis, and the Iraqis the worst), it has averaged about 2.0, or using the same manner of comparison, 100 Israelis in combat units have been comparable in performance to about 200 Arabs. (In 1973 the Israeli-Iraqi comparison was about 3.4.) It must be stressed that this does not mean that the Israelis were stronger, smarter, braver, or more highly motivated than the Arabs. It is simply that they were better organized, and more professional in the use of their weapons and equipment.

It appears that, as a result of their experience in the eight-year Iran-Iraq War, the Iraqis have improved greatly since their abysmal performance in 1973. There is considerable evidence that this improvement has probably brought them up about to the standard of the Egyptian Army. Thus the Israeli-Iraqi CEV can be estimated as now about 2.0. There is little doubt that the quality of the US Army today is probably on a par with that of the Israeli Army. Thus the US-Iraqi CEV is also probably about 2.0.

APPENDIX F

Logistics in Historical Perspective

In ancient and medieval armies, before the advent of gunpowder, logistics was a comparatively simple affair, but far from easy to manage. Troops had to be provided with clothing, armor, weapons, and food. The army's animals, including the mounts for the cavalry as well as transport animals for mobility, also needed food and water, as well as equipment like horseshoes, saddles, and harnesses, and people to tend them. The greatest problem faced by the logisticians of these pre-modern armies was keeping everybody and every animal fed. Generally speaking, a man ate about three pounds of food per day (mostly bread or other grain products), and a horse needed ten pounds of grain and again that much grass, fodder, or silage. For a modest army of 20,000 men and 5,000 horses, this amounted to a daily consumption of 85 tons of edibles, including 50 tons for the horses. Along with the necessity of gathering fodder or finding pasturage for the horses, soldiers usually supplemented their grain or bread rations with meat, cheese, vegetables, and fruit gathered on the march.

In addition to food and fodder, armies were also burdened with other items, like tentage, officers' baggage, field forges, and engineering tools and supplies. The men and animals needed to pull and drive the wagons carrying this material had to be fed as well, and so imposed further supply burdens.

Although soldiers commonly carried on their persons enough food for a few days (mostly flour or biscuit supplemented by beans or salt meat), horses generally depended on grazing from local pastures. For purely logistical reasons, an army regularly had to find fresh sources of fodder regularly, or it faced serious problems keeping its animals healthy and able to work. Sieges caused particular problems, since the besieging army was obliged to camp in one area for an extended period. Sometimes sieges had to be abandoned because the besieging force could no longer feed itself. Military strategies which have sometimes puzzled modern historians, such as the route of Alexander the Great's army in the Persian Empire, make sense only when the problems of logistics are taken into account.

Gunpowder and the Rise of Magazines

The introduction of firearms brought another factor to the logistics equation, although at first a relatively minor one. Ammunition expenditure in armies equipped with black powder firearms was very small compared to modern forces. Even as late as the Franco-Prussian War of 1870-1871, the average Prussian soldier expended only 56 rifle rounds during the entire war. Frederick the Great, who made great use of his cannon, provided them with an average of 180 balls per tube for the usual campaign.

The development of the magazine system, undertaken both to cure the ruinous depradations of the armies of the Thirty Years' War—which had supplied themselves largely through forage and plunder of the countryside—but also to regularize the quality and issue process of supplies to troops, had notable effects on military operations. First, armies were linked to their magazines by lines of shuttling supply wagons, and so they were unable to go more than a few days' march from the nearest magazine. Second, enemy forces interdicting or severing the supply line to an army's magazines would force the army to suspend operations, to find another magazine, or to reopen communications with the original. Concern for secure supply lines promoted a war of maneuver, since it was possible to gain victory simply by placing an army astride the enemy's supply lines. The magazine system also made long-distance movements more difficult, and the necessity of gathering supplies ahead of the army during a move such as Marlborough's march to the Danube in 1704 makes such a move all the more remarkable.

The French armies of the Revolutionary and Napoleonic Wars discarded the magazine system to a degree, although in the case of the armies of the First Republic this was a matter of necessity (they lacked the organizational capacity to establish and maintain magazine systems) rather than choice. Napoleon himself employed a mixture of magazine and foraging logistics, and the flexibility and ingenuity of the French logistics system provided Napoleon's armies with superior mobility. Even in the ultimately disastrous Russian campaign of 1812, the logistical difficulties of the Grande Armée were due less to any failure to prepare or procure supplies, than to the army's indiscipline, which reduced the quantity of supplies gained by foraging.

The American Civil War presaged a revolution in logistics. Possibly the most important of the contributions of the Industrial Revolution to the logistical dimensions of military art and generalship in that conflict were the telegraph and the radio. William Tecumseh Sherman, who has been called "the prophet of modern war," wrote in his *Personal Memoirs* about "the value of the magnetic telegraph [and] the value of railways" as follows: "Hardly a day intervened when General Grant did not know the exact state of facts with me more than fifteen hundred miles away as the wires ran. . . . The Atlanta campaign would simply have been impossible without the use of the railroads from Louisville to . . . Atlanta. . . ." (v. 2, 398).

New weapons in the Civil War, including the breech-loading, rapid-firing carbine and rifle, greatly increased the range and accuracy of firepower—and also increased ammunition consumption. These and other new developments added weightily to the logistic foundation, the "tooth-to-tail" ratio that compares the number of soldiers performing rear-area support services to the numbers supported in the combat arms: infantry, cavalry (now armor), and artillery. (The largest ratio today in Operation "Desert Shield" is 16:1 to support each pilot in an Air Force tactical aircraft or an Army Apache helicopter; the average ratio of service soldiers to combat soldiers is 8:1.)

Although railroads could transport vast quantities of supplies quickly and easily, distribution and transport to the troops in the field from the railhead still presented a serious problem. The Prussian army of 1870-1871, with supply and transport services that were extremely well-organized by contemporary standards, was unable to support its forces in the field as intended. Consequently, the army was compelled to forage for food and fodder; during the siege of Paris, Prussian troops had to harvest crops to feed themselves.

The Logistical Burden of Industrial War

The full effect of the Industrial Revolution on warfare did not make itself felt until World War I, and that conflict represents the true advent of modern logistics. For instance, while the average Prussian artillery piece in the six months of the 1870-1871 war with France expended only 199 rounds, the introduction of modern quick-firing guns meant that during the first six weeks of World War I, the typical German field gun fired nearly 1,000 rounds, a twenty-fold increase in rate of expenditure over 44 years. Yet it was not uncommon in World War II for a gun to fire 200 rounds per day for extended periods, another ten-fold increase. Small-arms ammunition expenditure rose in a similar fashion, from 56 rounds in six months (1870-1871) to 280 rounds in six weeks (1914).

These trends continued during the course of the war. A typical British division of 18,000 men in 1914 required 27 wagon-loads of all classes of supplies per day, but within two years each reorganized 14,000-man British division required 50 wagon-loads, 30 of these bearing combat supplies. The full logistical effect of modern weapons made it impossible for an army to carry with it all the munitions it would need for a campaign; it was essential that it be resupplied continually. When the burden of fuel, lubricants, and spare parts for motor vehicles are added to the increasing quantities of munitions consumed in modern combat, one begins to form an idea of the dimensions of supply required by a modern field army.

In the old days, an army could feed itself as long as it kept moving into new areas which had not been foraged over, and it encountered difficulties only when it tried to remain in place. By 1914, the opposite was true: an army could be assuredly supplied so long as it remained in one place near its supply sources; when it moved away, problems quickly ensued.

During World War I the new logistical constraints severely limited the mobility of armies, which could be readily supplied only near railroads or waterways. The growing use of motor vehicles only slightly mitigated these constraints during that war. Only the wide-scale employment of motor vehicles after World War I freed armies to move away from their sources of supply or from the railheads. However, this new freedom of movement carried with it the cost of imposing a further logistical burden: that of fuel, lubricants, and spare parts for vehicles. Even in World War II motor transport was not universally employed. In fact, only the British and American armies were primarily motorized; the German and Soviet armies were only partly motorized, with most of their infantry transport still horse-drawn. The Japanese army was almost entirely dependent on animal transport in 1941, although most divisions had acquired some motor transport by 1944-1945.

Another innovation in World War II was the use of air transport. Again, only the Anglo-American Allies had the material capability to employ this on a large scale. Particularly interesting was the reconquest of Burma in 1944-1945 by the Chinese-American army of General Joseph W. Stilwell in the north, and British General (later Field Marshal) Sir William J. Slim's Fourteenth Army in central and southern Burma. Stilwell's and Slim's troops entered Burma on the ground, but most of the supplies were flown in, freeing their forces from dependence on Burma's woefully inadequate road and railroad systems. Some of Stilwell's units were totally dependent on air supply for months on end. The Japanese defenders, who were reckoning on logistical difficulties to halt the Allied offensives and provide the opportunity for counterattack, were sorely disappointed. Their first counteroffensive failed at Imphal and Kohima in early 1944, and a small Japanese army literally starved to death because it was unable to get enough food over jungled mountain trails during the monsoon rains. When the final Japanese counterattack came, after the British reached the plains of central Burma, in March 1945, it was shredded by British and Indian tank and motorized forces.

As military units and their equipment have grown more complex, the resulting logistical requirements have grown as well. Not only does new equipment require new supplies and new spare parts to keep it operating, but also new specialists to run it, maintain it, and to fix it when it inevitably breaks down. For instance, by the end of the Korean War, standard US daily consumption estimates ran to 300 to 640 tons for infantry division, and 350 to 700 tons for armored divisions, depending on their activity. Even in reserve and out of combat, an armored division consumed 136 tons and an infantry division 111 tons each day. Consumption rates have risen still higher since then, as those 1953-era units were equipped mostly like their World War II predecessors, while modern US Army units have their own helicopters, more armored equipment, and vastly more extensive communications and radar equipment.

SELECT BIBLIOGRAPHY

Chubin, Shahran, and Charles Tripp. *Iran and Iraq at War*. London: I.B. Tauris and Co., Ltd., 1988.

Cordesman, Anthony H. *The Iran-Iraq War and Western Security, 1984-1987*. London: Jane's Publishing Co. for the Royal United Services Institute (RUSI), 1987.

Crichton-Stuart, Captain M. "The Story of a Long-Range Desert Patrol." *Army Quarterly* 47 (October 1943) and 48 (January 1944).

Dudgeon, Anthony G. *The War That Never Was*. Unpublished book ms. supplied with kind permission of Air Vice Marshal Dudgeon.

Dupuy, R. Ernest, and Trevor N. Dupuy. *The Encyclopedia of Military History*. Second revised edition. New York: Harper and Row, 1986.

Dupuy, Trevor N. *Attrition: Forecasting Battle Casualties and Equipment Losses in Modern War*. McLean, Va.: HERO Books, 1990.

———. *Numbers, Prediction, and War: Using History to Evaluate Combat Factors and Predict the Outcome of Battles*. New York: Bobbs-Merrill Co., Inc., 1979.

Engels, Donald W. *Alexander the Great and the Logistics of the Macedonian Army*. Berkeley: University of California Press, 1980.

Esposito, Col. Vincent J. *West Point Atlas of American Wars*, Vol. II, *1900-1953*. New York: Frederick A. Praeger, 1959.

Helms, C. M. *Iraq: Eastern Flank of the Arab World*. Washington, D.C.: Brookings Institution, 1984.

International Institute for Strategic Studies (IISS). *The Military Balance, 1990-1991*. London: Brassey's for IISS, 1990.

Isby, David C. and Charles T. Kamps, Jr. *Armies of NATO's Central Front*. Jane's Publishing Co., London, 1985.

Jenner, Bob, and List, David. *The Long-Range Desert Group*. London: Osprey Publishing, 1983.

Miller, David, and Christopher F. Foss. *Modern Land Combat*. New York: Portland House, 1987.

Naff, Thomas (ed.). *Gulf Security and the Iran-Iraq War*. Washington, D.C.: National Defense University Press, 1985.

Pelletiere, Stephen C., Johnson, Douglas V, II, and Rosenberger, Leif R. *Iraqi Power and US Security in the Middle East*. Carlisle, Pa.: Strategic Studies Institute, US Army War College, 1990.

Sirriyeh, H. "Development of the Iraqi-Iranian Dispute, 1847-1975." *Journal of Contemporary History*, 20:3 (1985).

Syrett, David. "Long-Range Desert Operations." *Military Review* 54 (January 1984).

Toppe, Alfred. *Desert Warfare: The German Experience in World War II*. Tr. and ed. H. Heitman. N.p.: Historical Division, European Command, n.d.

US Army. FM 63-3J, *Combat Service Support Operations--Corps*. Washington, D.C.: US Department of the Army, August 1985.

_____. FM 100-5-1, *Operational Terms and Symbols*. Washington, D.C.: US Department of the Army, October 1985.

_____. FM 101-10, *Staff Officer's Field Manual*. Washington, D.C.: US Department of the Army, July 1953.

_____. FM 101-10-1/2, *Staff Officer's Field Manual: Organizational, Technical, and Logistical Data Planning Factors* (Vol. 2). Washington, D.C.: US Department of the Army, October 1987.

Van Creveld, Martin. *Supplying War: Logistics from Wallenstein to Patton*. Cambridge: Cambridge University Press, 1980

INDEX

air transport 159-60
airborne or parachute operations 75, 78
airborne early warning (AEW) aircraft (AWACS, E-2C Hawkeye) 57
airmobile or air assault operations 48, 50, 51, 65, 73, 75, 78, 101, 102
airpower 48, 53, 54
amphibious assault 48, 71, 73, 75
Anglo-French provisional corps 31, 41, 62, 71, 74, 78
Anglo-Iraqi Conflict (1941) 4-6, 117-19
Arab Israeli Wars:
    First (1948-49) 10, 14, 118
    Third (1967) 10, 14, 43, 49, 54, 119
    Fourth (1973) 13, 37, 39, 49, 54, 119
Arab League 8, 14, 80, 83, 118-22

Ba'ath Party 9-11, 14, 17, 96, 122, 123
Baghdad 5, 6, 8, 26, 58, 68, 69, 73, 95, 128
Bahrain 4, 8, 12, 120, 121
Basra 5, 6, 15, 37, 58, 76, 126, 127
blockade or siege 47, 48, 51, 96-100, 102
Britain (United Kingdom) 3, 4-6, 8, 99, 116-20
    British forces 31, 41, 42, 71, 74, 134, 135, 145-47
    RAF (Royal Air Force) 5, 6, 58, 117, 147
    1st Armored Division 71, 145, 146
Bubiyan Island 45, 51, 77, 98, 124, 125

"carpet bombing" (and Arclight raids in Vietnam) 63-66
casualties (forecasted) 104-06
CEV (realtive combat effectiveness), see also TNDM 42, 153-55
Chamberlain, Neville 1, 19
chemical and biological weapons and warfare 19, 20, 45, 49, 50, 55, 56, 58
climate and weather 36, 123, 127-32
combined (multinational) operations 28
command 27-33, 50, 55
Command and General Staff College; see US Army
cruise missiles 57

Egypt 13, 25, 36, 116, 118-21
    Egyptian forces 31, 42, 43, 54, 71, 74, 136-37, 145, 147
    3d Armd. Div. 147
    3d Mech. Div. 147
envelopment
    single or close 46-51, 68, 70, 72, 101
    wide or strategic, turning movement 46-49, 51, 75-78, 101
estimate/appreciation of situation 35-53
Euphrates River 1, 6, 8

Faw (Fao) 15, 126-27
Fertile Crescent 1, 8
field fortifications 59, 63, 65-67

France 10, 98
  French forces 31, 40, 41, 71, 74, 136, 137, 145-47
    6th Lt. Armd Div. 71, 145, 146
frontal assault 45-50, 62-69, 102

Gulf Cooperation Council (GCC) 12, 14, 121
Gulf States (see also individual countries) 8, 10, 12, 13, 116-22, 145

Habbaniya 5, 6, 119
helicopters 50, 58, 59, 86, 89, 132

IFF (identification friend-or-foe) 44
Iran 3, 10, 12, 14, 15, 17, 111, 121, 122
Iran-Iraq (Gulf) War 14-17, 59, 65, 121, 127
Iraq 1, 3, 4, 6-11, 14-17, 22, 26-28, 36, 43, 54, 57-59, 74, 80, 81, 95, 96, 105, 106, 117-22, 126-34
  Iraqi forces (general) 17, 34, 37-39, 49, 80-84, 92, 95, 138, 139, 147-51
  Iraqi Army 5-6, 15, 37-9, 41, 72, 147-51
    Republican Guard 39, 50, 69,
    South Kuwait Defense Force 38, 72, 75, 77, 101, 147-51
    South Iraq Defense Force 39, 72, 77, 101, 151
    Coastal Defense Force 38, 39, 151
    Local Reserve Force 39, 73, 75, 78, 151
    Strategic Reserve (Basra) Force 39, 75, 151
  Iraqi Air Force 38, 41, 47, 48, 55, 56, 146, 149, 151
  Iraqi Navy 149
Israel 10, 13, 28, 36, 44, 80-83, 110, 118
  Israeli Air Force (IAF) 44, 55
  Israeli Army (IDF) 83, 130
Italy 147, 149
  Italian invasion of Ethiopia (1935-36) 21, 22, 24

Japanese conquest of Manchuria (1931-32) 21, 22, 24
joint (multi-service) operations 28
Jordan 11, 16, 17, 80-84, 98, 102, 114, 118, 128
  Jordanian forces 43, 82-83

Khuzistan/Arabistan 4, 12
Kurds 9-10, 44, 121
Kuwait 2-4, 8, 10, 14-17, 22, 24, 25, 36-39, 42, 55, 58, 59, 66, 69, 72, 74, 76, 96, 98, 99, 108, 115-21, 125-27

Kuwaiti forces 31, 59, 135, 136, 145

League of Nations 4, 7, 10, 14, 21, 22
Leavenworth, Kansas, Fort 49, 70, 71, 78
logistics 4-6, 44, 49, 55, 60, 85-95, 127-30, 156-60
  distribution 86-87, 89, 90-93, 159, 160
  production 86, 89, 91
  supply 55, 85, 86-89, 95-97, 98, 156-60
  support commands (US Army) 86

Mesopotamia 8, 129
motor transport and motor vehicles 85, 86, 128-29, 159-60
Munich Agreement (1938) vi, 19, 20, 23
NATO 12, 28
nuclear weapons 55, 110

oil 4, 10, 12-13, 15, 21, 24, 43, 117, 118, 122
  oil reserves 13
Oman 4, 8, 12, 120
operations (operational art) 34, 35, 46-49
Operation "Bulldozer" 50, 50, 62-9, 71, 101, 103, 104, 106
  "Colorado Springs" 49, 53-60, 67, 78, 83, 95, 97, 100, 102-104
  "Desert Shield" 16, 32
  "Leavenworth" 51, 52, 70-73, 74-78, 97, 99, 101, 104-06
  "RazzleDazzle" 51, 74-9, 97, 101, 104-06
  "Salah-al-Din" 61, 68, 69, 73, 78, 79
  "Siege" 52, 95-99, 101, 104-106
Ottoman Empire 2-4, 6, 9, 54, 116, 117

Palestine and Palestinians 2, 4, 14, 111-15, 117, 118
Persian Gulf i, 2, 6, 13, 18, 26, 57, 63, 74, 97, 116-22
PLO 113-17
posture 42, 153, 155

Qatar 4, 8, 10, 120, 121

Ramadan 37

sabkhas (defined) 123-24
Saddam Hussein ii, 10, 14-20, 24-25, 34 58, 80-82, 96, 108-15
sand and dust (environmental effects of) 126-132
Saudi Arabia 7-8, 12, 24-25, 27-29, 44, 98, 114, 116-21, 125, 137, 138

Saudi Forces 31, 42, 60, 71, 136, 145, 147
Saudi I Corps 69, 145
Scud-B ballistic missiles 14, 45, 58, 80
stealth aircraft (F-117A) 58
strategy 34, 35
Syria 10-11, 17, 25, 27, 36, 117-119
  Syrian Forces 27, 31, 42, 43, 53, 71, 72, 96, 101, 137-38, 140, 145-47, 150
    9th Armd. Division 145, 150

tactics 34, 35, 46
  Iraqi defensive tactics 65-67
  UN forces offensive tactics 46, 47, 67, 68
terrain 4, 123-127
Tigris River 1, 8
Tigris-Euphrates river system 4, 59, 123, 126
TNDM (Tactical Numerical Deterministic Model) 102, 103, 153-55
Turkey 25, 119
  Turkish forces 27, 50, 96, 101

United Arab Emirates (Trucial Coast) 4, 12, 120, 121
United Nations 10, 22-26, 112, 113
  UN Allies 27-32, 34, 41, 42, 54, 80, 106, 107
  UN Security Council Resolution 22, 36, 83
  UN Forces (general) 27-32, 34, 35, 42, 62, 70, 74, 79, 83, 92, 95, 98, 99, 100, 106, 145, 146
    UN Air Forces 49, 50, 94, 95
    Eastern Army 29-31, 36, 45, 49, 51, 60-62, 72, 74, 76, 95, 101
    Western Army 29-31, 46, 49, 61, 74, 76, 95, 101
United States i, 7, 22, 24, 27, 33, 80-84, 97, 99, 114-16

US Army 31, 62-9, 100, 140-44
  Seventh Army 96
  Command and General Staff College 50, 70, 71, 78
  VII Corps 39, 62, 69, 71, 73, 76, 78, 79 143
  XVIII Abn Corps 39, 62, 70, 71, 73, 74, 76, 141
    1st Cav Div. 72, 75, 141, 142
    1st Armd Div 75, 77, 78, 141, 142
    1st Mech. Inf. Div. 72, 140-43
    2d Armd. Div. 73, 141-43
    3d Armd. Div. 72, 78, 143
    24th Mech. Inf. Div. 72, 140, 141
    82d Airborne Div. 73, 76, 84, 141, 144
    101st Abn. Div. (Air Assault) 73, 76, 84, 141, 142
US Air Force 40, 41, 53-60, 64-66, 91, 104, 143, 144
US Marine Corps 39, 49, 50, 62, 68, 74, 89, 130, 144, 148
  1st MEF 71, 72, 74, 78, 143
  1st Division 73, 77, 143, 144
  2d Division 73, 76, 78, 143, 144
  4th MEB 73, 143
  5th MEB 78, 143
  MEB (general) 40, 61, 73, 77, 78, 101
US Navy 41, 57, 91, 122, 144
  aircraft carriers 57, 91, 144
US forces (general) 27-32, 42, 64, 106, 140-44

wadis (defined) 123, 124
weather: see climate and weather